KB076795

김이홍 Leehong Kim

문주호 Jooho Moon
임지환 Jihwan Lim
조성현 Sunghyeon Cho

남정민 Jungmin Nam

2018

젊은 건축가: 경계의 가치를 묻다

김이홍 문주호 임지환 조성현 남정민

펴낸날: 2018년 10월 5일 초판

글쓴이:
김이홍 문주호 임지환 조성현 남정민
존 홍 이민아 김현섭

펴낸이: 김옥철
주간: 문지숙
기획편집: 박성진
진행: 임선희
편집 도움: 하명란 김나래 서하나
번역: 홍근호, 이영주
디자인: 유윤석
사진: 신경섭, 이호경, 배지훈, 노기훈, PACE스튜디오, 송유섭
제작 도움: 박민수
커뮤니케이션: 이지은 박지선
영업관리: 김헌준 강소현

인쇄: 스크린그래픽
제책: SM북

펴낸곳: (주)안그라픽스
우10881 경기도 파주시 회동길 125-15
전화 031.955.7766(편집) 031.955.7755(마케팅)
팩스 031.955.7744(팩스)
이메일 agdesign@ag.co.kr
웹사이트 www.agbook.co.kr
등록번호 제2-236(1975.7.7)

제11회 젊은건축가상 2018
주최: 문화체육관광부
주관: 새건축사협의회 한국건축가협회 한국여성건축가협회
후원: 국민체육진흥공단

이 도서의 국립중앙도서관 출판예정도서목록(CIP)은 서지정보
유통지원시스템 홈페이지(seoji.nl.go.kr)와 국가자료공동목록시스템
(www.nl.go.kr/kolisnet)에서 이용하실 수 있습니다.
CIP제어번호: CIP2018030010

ISBN 978.89.7059.980.9 (93540)

Young Architect: Asking the Value of Boundaries

Leehong Kim, Jooho Moon, Jihwan Lim, Sunghyeon Cho, Jungmin Nam

Publication Date: October 5, 2018

Authors
Leehong Kim, Jooho Moon, Jihwan Lim, Sunghyeon Cho, Jungmin Nam,
John Hong, Minah Lee, Hyon-Sob Kim

President: Okchyul Kim
Chief Editor: Jisook Moon
Planning & Editing: Sungjin Park
Assistant: Sunhee Lim
Proofreading: Myungran Ha, Narae Kim, Hana Sur
Translation: Keunho Hong, Yeongju Lee
Design: Yoonseok Yoo
Photograph: Kyungsub Shin, Hogyeong Lee, Jihun Bae,
    Gihun Noh, PACE Studio, Yousub Song
Production support: Minsoo Park
Communication: Jieun Lee, Jisun Park
Customer Service: Heonjun Kim, Sohyun Kang

Printing: Screen Graphic
Binding: SM Book

Publisher: Ahn Graphics Ltd.
125-15, Hoedong-gil, Paju-si,
Gyeonggi-do 10881, Korea
tel +82.31.955.7755
fax +82.31.955.7744
email agdesign@ag.co.kr
www.agbook.co.kr

The 11th Korean Young Architect Award 2018
Presented by The Ministry of Culture Sports and Tourism
Organized by Korea Architects Institute, Korean Institute of Architects,
Korean Institute of Female Architects
Sponsored by Korea Sports Promotion Foundation

A CIP catalogue record for this book is available from
the National Library of Korea, Seoul, Republic of Korea.

CIP Code: CIP2018030010

본 사업은 국민체육진흥기금 후원을 받아 시행하는 사업입니다

젊은 건축가
경계의 가치를 묻다

Young Architect:
Asking the Value of Boundaries

안그라픽스

Leehong Kim

Jooho Moon Jihwan Lim Sunghyeon Cho

Jungmin Nam

프롤로그

건축적 사유와 실천의 복기

박성진 · 책임에디터, 사이트 앤 페이지 디렉터

젊은건축가상(KYAA)은 '국내외 건축사 자격증을 취득한 만 45세 이하'의 느슨한 조건과 '창의적이고 진취적인' 동시에 '독창적인 사고와 새로운 방향'이라는 까다로운 시험대를 통과한 선택된 소수에게 주어지는 영예이다. 올해 지원한 31팀 가운데 선정된 3팀의 수상자들에게 먼저 축하의 말을 전한다.

정도의 차이는 있지만 그들은 기성 건축가들과 다른 행보를 보이며 한국 건축계의 지평을 넓혀주는 새로운 모색과 동력으로 이미 자리 잡았다. 그러니 내가 그들의 건축적 성취에 딴지를 걸 일도 없고, 심사위원들이 꼼꼼하게 따져본 그들의 자격을 의심할 일도 없다. 그렇다고 주례사처럼 심사 총평을 다시 읊을 일도 아니다. 다만 나는 이 책의 서문에서 그들에게 다른 질문과 메시지를 던지고 싶다. 젊은 건축가에게 책은 무엇인가? 수상자에게 이 출판의 의의는 무엇일까?

그들 대부분 출판 경험이 없다. 책이라는 매체와 편집저작물을 통해 자신의 건축을 정리하고 설명해본 경험이 전무하다. 가끔 잡지라는 짧은 호흡의 지면을 통해서만 출판을 찔끔찔끔 경험했을 것이다. 어엿한 지은이로서 이번처럼 출판의 기회를 갖는다는 것은 젊은건축가상이 마련해주는 또 다른 세계로의 진입이다. 솔직히 그들은 글로 자신의 생각을 적는 데 서툴고, 문장은 거칠고 중언부언하며, 주어와 동사는 이래저래 따로 노는 등 아직 출판에서는 미숙한 티를 벗지 못한 초짜들이다. 책이라는 한정된 크기와 종이의 물성, 그리고 페이지들의 내러티브를 통해 자신의 작품들을 어떻게 보여주어야 할지 확신이 없다. 그래서 경험 있는 기성 건축가와 책을 준비할 때보다 그들과 이 책을 준비하면서 훨씬 손이 많이 갔고, 세세한 설명과 주문을 반복하는 지난한 노력이 수반되었다. 고백하자면 그들이 보낸 원고와 자료를 살피고 정리하면서 솔직히 울화가 치민 적이 몇 차례 있다.

그럼에도 나는 확신한다. 수상의 소식과 영광은 그들을 새로운 반열에 올리지만, 그들에게 건축에 대한 새로운 탐구와 고민의 시간을 열어주는 건 아니다. 앞서 말한 새로운 건축적 탐침(探針)은 수상 이후 바로 이 책을 준비하면서 시작되는 경험들이다. 수상자들은 책을 준비하면서 다시 한 번 작업들을 테이블 위에 올려놓고 도시적, 조형적, 사회적 가치와 의미를 따져가며 개념적으로 수십 번 해체와 재구성을 반복했을 것이다. 자신의 건축물이 이 도시와 사회와 이야기 나누고 싶었던 주제가 무엇인지, 그 수많았던 사유와 실천을 복기하며 사진을 한 장 한 장 고르고, 도면 위 선을 지웠다 그리기를 반복했을 것이다. 경우에 따라 그 과정에서 없던 도면과 다이어그램도 새로 그리고, 심지어 없던 개념까지 만들어내는 A/S가 있었을 것이다.

특히 글을 쓸 때는 더욱 그렇다. 오늘날 젊은 건축가들은 글쓰기 훈련이 거의 되어 있지 않다. 그들이 건축의 복합성과 자의성을 오로지 그들의 선험적 감각과 판단의 영역으로만 남겨둔다면 우리 건축계의 담론 형성은 풍요를 얻지 못하고 소통의 난맥을 겪을 것이다. 젊은건축가상은 일개 기업이 아니라 공공이 주는 상이다. 자신의 건축을 다른 이에게 합리적으로 설명하고 공공과 소통할 수 있어야 수상의 취지가 살아난다. 간혹 건축가는 글을 쓰면서 자신도 몰랐던 작업의 의도와 의미를 발견하고 만들어내곤 한다. 그것이 지나친 수사와 관념으로 물들 위험이 있지만, 그래도 자가당착의 늪에 빠지지만 않는다면 그것도 그리 나쁘진 않다. 진정성 있게 한 쪽의 글을 쓰려면 열 쪽의 다른 글을 읽어야 하고, 세 쪽의 글을 일단 두서없이 써봐야 한다.

경계없는작업실에 대해 글을 쓴 이민아는 건축가의 글과 말에 대해 비평하는데, "결국 건축가는 자신이 한 일에 대해 글을 쓰고 건축가가 선택하는 어휘의 비약은 그들의 고뇌가 어느 지점인지를 드러낸다. 건축적 글쓰기는 매우 위선적일 수 있는 도구이면서 유일하게 순간마다 자신을 성찰하게 만드는 고해 과정이기도 하다."라고 말한다.

그래서 올해 수상자들에게 빠듯한 일정에도 어려운 주문을 건넸다. 기존에 있던 글을 정리하거나

조합하지 말고 짧더라도 자신의 건축에 관한 새로운 글을 써보자고. 그들이 건축적 사유를 형성해가는 데 지대한 영향을 끼쳤던 현상이나 인물, 대상, 경험 등에 대해 파편적인 글쓰기를 주문했다. 그리 길지도 않은 그 글을 건축가들은 약속했던 마감일을 어기고 또 어겨가며, 전전긍긍 끌어안고 고민하며 써나갔다. 나는 이 과정에서 수상의 기쁨과는 전혀 다른 성장통의 고역을 경험하며 그들 건축에 대한 작은 발견이 있었을 거라고 생각한다. 그리고 그 발견은 본인의 글뿐 아니라 비평가로 합류한 존 홍, 이민아, 김현섭의 원고를 통해서도 발견했을 것이다.

한국 건축계에서 그간 출판은 건축가라는 위상을 확립해주고 이를 공고히 하는 수단으로써 유효했다. 건축가의 작업들을 열거한 작품집이든, 건축적 사유를 묶어 펴낸 자기성찰의 에세이집이든, 우리 도시와 건축에 대한 폭넓은 교양서든, 건축가는 계속 글을 쓰고 소통해왔다.

그런 측면에서 나는 젊은 건축가에게 주문하고 싶다. 앞으로 좀 더 쓰자. 누군가에게 읽혀지고 자기를 포장하기 위한 글이 아니라 자신에게 되뇌고 자신에게 다시 읽혀질 글을 건축가로서 써가자는 말이다. 자신의 건축과 이 도시에 대한 글쓰기를 주문하고 싶다. 어느 순간 그 글은 큰 줄거리를 갖고 그들의 건축에 새로운 지침으로 다시 돌아올 것이다. 이 책은 앞으로 그들이 만들어갈 책의 I막 I장, 서막일 뿐이다.

Restoring Architectural Contemplations and Activities

by Sungjin Park (editor / director, Site & Page)

The Korean Young Architect Award is an honor given to a small number of individuals who have not only passed the relatively easy condition of 'having an architect's license either from Korea or abroad and being under age of 45' but also the selective process that looks for 'a creative and progressive mind' and 'a unique and novel thought direction'. First, I'd like to congratulate the three awardees who were selected as winners out of the 31 participating teams this year. While they all differ to some degree, they all display a distinguished path from the regular architects and have already settled down as a new face and energy that expand the horizon of the architectural realm in Korea. Because of this, there is no need for me to write about their architectural success or doubt their qualities and the judges' decision. There's also no need for me to repeat what has already been covered in the general review. Instead, I only wish to address this question and a message to the three winners in this prologue: what is a book to a young architect? What meaning does this publication hold for the winners?

Most of them do not have experience in publishing. They have no experience of organizing and explaining their architecture on a book and a editorial writing. Perhaps, they may have had a brief touch with publishing through a magazine. This opportunity to publish as an author of one's own work would have been like an entrance to a whole new world led by the Korean Young Architect Award.

Honestly speaking, with their unpracticed textual expressions, unpolished and repetitive sentences, and unmatched subject nouns and verbs, they are still green when it comes to publishing. They have no idea on how to present their works within the limits of a book, paper, and its pages. Because of this, much more editorial work that required tedious explanations and requests was needed to prepare this book than when editing books by other more experienced architects. To admit, there were a couple of times when I felt exasperated as I was looking through their manuscripts and materials.

Regardless, I am confident. The award and its honor may bring them up to a new level, but it does not provide them with a time for new kinds of exploration or contemplation. This new kind of architectural probing refers to the experiences that come about as they prepare this book after the award ceremony. In other words, to prepare for this book, the awardees must have placed their works back on their table, ruminated upon their urban, sculptural, and social value and significance, and had them conceptually deconstructed and rebuilt numerous times. They must have redrawn their lines again and again while carefully choosing their photos as they course back to their countless contemplations and activities to remind themselves of the theme that they wanted to demonstrate in regards to how their architecture engages with the city and the society. In some cases, they might have redrawn floor plans and diagrams or even create concepts that they did not have originally before the publication itself.

This is especially so when it comes to writing. Most of the young architects these days are not well-trained in writing. If they were to limit the complexity and autonomy of architecture solely to their theoretical intuitions and judgments, it would impoverish the discussions in architecture and create communicative difficulties. The Korean Young Architect Award is an honor that is not given by an individual company but by the public. The aim of the award is fulfilled only when one is able to rationally explain one's own work to others and communicate with the public. Sometimes, through their writings, architects come to discover the purpose and meaning of their projects that they were not aware of before. While the danger of falling into excessive rhetoric and idealism is something to worry about, it is not too bad of a thing as long as one keeps away from the pitfall of self-contradiction. To produce just 1 page of meaningful text, one first has to read 10 pages of other texts and spend time writing aimlessly for 3 pages.

Minah Lee, who wrote on Boundless Architects, observes as such regarding the prose and sayings of architects: "Ultimately, these kinds of reflective writings and the vocabular leap that architects choose and write reveal where they are in their contemplation. While architectural writings can be quite hypocritical, it is also a confession that can make one self-reflect time to time."

Despite their tight schedules, the awardees of this year were given a difficult task: that is, to stay away from their past written materials and to produce a new text on their architecture regardless of its length. They were asked to write fragmentary essays on the phenomena, individuals, objects, or experiences that influenced them significantly in forming their architectural worldview. For these texts that weren't that long, the architects struggled strenuously and they repeatedly went past their deadlines. Through this arduous process that is far apart from the joyful mood of an award ceremony, I think that these architects have made a certain discovery in regards to their architecture. Also, through their contributions, I believe that a similar kind of discovery must have happened for the critics—i.e., Hyon-Sob Kim, Minah Lee, and John Hong—as well.

In Korean architectural realm, publication has been the means to publicly establish and secure the position of an architect. Whether in the form of a work series that lays out one's works, or a collection of essays that contains one's architectural reflections, or a general introductory book that discusses the city and architecture, architects have continued to write and communicate through texts.

In that aspect, there is something I'd like to demand from the young architects. Let's write more. By this I don't mean to write something to promote oneself to others, but to write something that one can meditate and self-reflect upon as an architect. I'd like to ask the young architects to write about their own architecture and this city. One day, that text will come back to them as a new and enriched architectural principle. This book is merely the first scene of the first act of their book—it is merely the beginning of a prelude.

Translated by Keunho Hong

김이홍

홍익대학교
김이홍 아키텍츠

Leehong Kim

Hongik University
Leehong Kim Architects

김이홍
현재 홍익대학교 건축도시대학원
조교수이자 김이홍 아키텍츠의
대표이다. 연세대학교 건축공학과와
하버드대학교 GSD(Graduate
School of Design)를 졸업한 후
7년간 삼우종합건축사사무소와
스티븐홀아키텍츠에서 실무를
익혔다. 김이홍 아키텍츠를 연 뒤
한국과 미국을 오가며 다양한
프로젝트를 진행하고 있다.
2009 광주디자인비엔날레와
아모레퍼시픽미술관의 APMAP
2016에 참여했다. 미국건축사와
LEED(Leadership in Energy and
Environmental Design) AP이며,
서울시 공공건축가로 활동하고 있다.

Leehong Kim
is a practitioner (principal, Leehong
Kim Architects) as well as an
educator (assistant professor,
Hongik University). He graduated
from Yonsei University and the
Harvard Graduate School of Design.
His work experiences include Samoo
Architects & Engineers and Steven
Holl Architects. Since opening of
Leehong Kim Architects, he has
led various projects in both Korea
and U.S. He participated in the
Gwangju Design Biennale 2009 and
APMAP 2016 of the Amorepacific
Museum of Art. He is a registered
architect in New York State and a
LEED accredited professional, and
currently serves as a Seoul Public
Architect.

에세이

개념과 구축의 경계

김이홍

김이홍

I

중학교 시절에는 목수가 되기를 꿈꿨다. 그때 직접 디자인하고 손수 만든 목가구 두 개가 아직도 내 방에 있다. 목수의 꿈을 이루지는 못했지만 건축가를 위한 시작점이자 지금의 나를 만든 중요한 계기가 목공 작업실이다.

나의 경력을 돌아보면 늘 목공실이나 모델실이 있고, 크래프트맨십(craftsmanship, 장인정신)을 중시하는 환경에서 학업과 실무를 했다. 대학원 시절, 르 코르뷔지에의 작품 중 유일하게 북미에 준공된 하버드대학교 카펜터센터(Carpenter Center)의 목공실에서 아르바이트를 했고, LA 프랭크 게리 사무실에서는 6개월 내내 건축모형을 만들었다. 나는 인턴이었지만 파트너들도 모형을 만져가며 디자인하는 환경이었기에 당연시했다. 그리고 뉴욕 스티븐홀아키텍츠도 모형 없이는 설계가 무의미한 사무실이다. 하나의 프로젝트를 두고 3년 동안 설계 단계별로 다양한 스케일의 모형을 만들고, 이를 설계에 반영하는 도구로 사용했다. 단순히 보여주는 모형이 아닌 실제 건물을 대변할 모형이기에 시공을 고려하고 선과 면 하나도 신중히 구현하고자 회사의 다양한 도구를 활용했다.

내가 느끼는 즐거움은 이것이다. 손과 머리가 따로 움직이는 것이 아닌 손과 머리로 동시에 해결해가는 과정. 컴퓨터 모니터 앞에서의 작업과는 달리 목공실 작업은 손과 머리를 동시에 사용하는, 차분히 몰두할 수 있는 힐링의 시간이기도 하다. 그 과정에서 궁극적으로 추구하는 바는 나의 사고를 보다 완성도 높은 결과물로 이끌어내는 것이다.

2

키 5cm 플라스틱 인형 18만여 개가 손바닥을 위로 하여
유리판을 받치고 있는 작품 'Floor'

지름 4mm 작은 원 20만여 개가 만드는
물방울무늬 가득한 벽지 'Who Am We?'

2.5×5cm 타원형의 미군 인식표 7만여 개를 모아 만든
2.15m 높이의 갑옷 'Some/One'

지인들의 사인을 수놓은 수만 개의 실을
한 지점에서 당기는 낙하산병 'Paratrooper-1'

수많은 개체가 반복된 수작업으로 이루어진 예술 작품은 큰 감동
을 준다. 특히 작은 개체들이 모여 예상하지 못한 더 높은 차원의
결과물을 만들고, 개인과 집단에 대해 고민하는 서도호의 작품을
보면 경이로움이 느껴진다. 정교한 노동을 통해 만들어진 작품은
시각적으로 아름답기까지 하다.

건축도 노동과 반복이 집적된 구축 과정을 거쳐 탄생한다. 뉴욕의
균일한 도시 조직 속에서 벽돌 한 장의 디테일로 차별화된 건물을
표현하고자 한 57E130 NY 콘도미니엄 프로젝트. 여기서 벽돌은 큰
고민거리였다. 멀리서 보면 벽돌 건물이지만, 가까이서 보면 서로
다른 다섯 개의 각도를 가진 총 1만 2,000여 개의 벽돌로 만들어진
파사드다. 이는 뉴욕의 콘텍스트(context)를 존중하면서도 차별화
된 디자인을 갖는다.

"쉽게 선뜻 접근하지만 발을 담글수록 투명한 레이어가 많아서 점
점 깊이 들어가게 되는 그런 작품을 하고 싶어요."라는 서도호 작
가. 반투명 은조사 천을 사용한 그의 작업에서 보이듯, 투영되어
작품이 존재하는 장소 너머까지도 볼 수 있다. 그 너머에 존재하는
것이 서도호 작가의 작업 과정, 곧 정교한 노동의 과정이지 않을까.
그 정교한 노동의 행위 자체가 하나의 작품으로 비춰진다.

3

건축에서 물질과 비물질을 고민할 때, 미국 예일대학교의 바이네키 고문서 도서관(Beinecke Rare Book & Manuscript Library)을 찾곤 한다. 사람을 위한 건축물인지 의심될 정도로 개구부가 대지 레벨에만 있고 나머지 5-6개 층의 높이는 대리석으로 둔탁하게 마감된 건축물이다. 광장의 석재 바닥 그리드에 맞춰 조성된 육면체 덩어리가 하나의 조각품 같다. 빈틈없이 불투명하고 정밀하게 재단된 외모를 갖고 있다. 하지만 그 안을 보는 순간 예상치 못한 반전이 펼쳐진다. 외부와 동일한 내부 파사드이지만, 오히려 투명함과 경쾌함이 호기심을 자극한다. 말로 설명할 수 없는 감동이 내부 공간에서 느껴진다.

공간은 눈에 보이지 않는 비물질적인 요소이며 건축의 본질이라 할 수 있다. 외부와 내부, 물질과 비물질의 본질이 다른 실체가 32mm 두께의 대리석 판으로 나뉜다. 외부에서 보았던 대리석 덩어리가 내부에서는 얇은 창호지처럼 느껴진다.

투명과 불투명을 0과 1의 스케일로 놓고 본다면, 0과 1은 건축가가 완벽히 구현 가능하다. 하지만 그 사이의 무수한 투명도는 인간이 설정한 바탕에 자연 요소가 더해져 마무리된다. 이 도서관은 빛의 강도에 따라서 내부 공간의 분위기와 모습이 시시각각 변한다. 강렬한 빛을 통해 대리석에 내재된 패턴과 밀도가 드러나고 예측하지 못한 분위기를 자아낸다. 물질적인 재료의 모습이 비물질적인 공간으로 전환되는 경계에서 우리들은 감탄한다. 건축의 물질과 비물질은 절대 32mm 정도의 아슬아슬한 차이가 아니다.

김이홍

4

0.83% 경사는 걸으면서도 인지할 수 없을 정도의 아주 완만한 각
도이다. 하지만 뉴욕이라는 거대도시에서 인상 깊게 다가온 순
간 중 하나가 0.83%의 경사를 발견한 순간이다. 물론 정확한 경
사도는 나중에 책을 통해서 알았지만, 180ft(54.864m)의 길이에
18in(45.72cm)의 높이가 상승하는 경사면이다.

뉴욕 링컨센터(Lincoln Center for the Performing Arts) 단지는
다양한 외부 공간으로 구성되어 있다. 외부 공간 중에서 3개의 공
연장 건물에 둘러싸인 광장은 단연 시민들이 즐겨 찾는 지점이다.
광장 정중앙의 분수대가 링컨센터의 주인공인 듯 매일 밤 화려한
조명을 받으며 물소리로 연주를 하고 물줄기로 발레 공연을 한다.

하지만 내가 더 주목하는 지점은 링컨센터 단지의 북측에 위치한
허스트플라자(Hearst Plaza)의 폴 밀스테인 풀(Paul Milstein
Pool)이라고 불리는 얕은 못이다. 밋밋한 석재 바닥 패턴의 공허
한 광장 한가운데에 지면과 유사한 레벨에 잔잔히 멈춰 있는 수면
은 규모(약 42.6×19.2m)를 제외하곤 크게 자신을 뽐내며 드러내
지 않는다. 오히려 가운데 자리 잡고 있는 영국 조각가 헨리 무어의
'Reclining Figure' 조각상의 캔버스로 보일 수도 있다. 하지만 광장
바닥 면이 0.83%의 경사면이라는 사실의 단서를 제공하는 못의 수
면은 더 묵묵히 그리고 고상하게 존재감을 드러낸다. 못의 한쪽은
광장보다 15cm가량 낮고, 반대쪽은 15cm가량 높으며, 중간 지점
에서 그 지점과 만나는 플라자와 동일한 레벨을 갖는다. 재료를 통
해 표현된 섬세함이 이 도시 프로젝트에서 느껴진다.

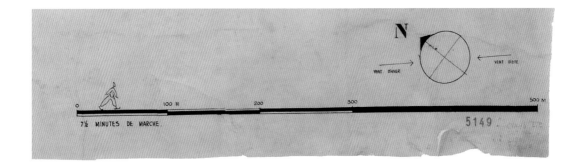

김이홍

5

2013년 6월 뉴욕 현대미술관(MoMA)에서 르 코르뷔지에의 대규모 회고전 〈르 코르뷔지에: 모던 랜드스케이프 아틀라스(Le Corbusier: An Atlas of Modern Landscapes)〉가 열렸다. 스케치, 도면, 모형, 사진 등 320여 점이 미술관 한 층을 가득 채웠지만, 그중에서도 뇌리에 가장 깊게 남은 것은 특정한 프로젝트나 도면, 모형도 아닌 인도 찬디가르 국회의사당(L'Assemblée Nationale de Chandigarh) 단지의 배치도 오른쪽 아래에 그려진 작은 스케일바이다.

그 스케일바는 통상적인 형식이지만, 스케일바 위에서 스케이트를 타는 사람의 역동적 모습과 '7 ½ MINUTES DE MARCHE (걸어서 7.5분 거리)'의 문구는 위트 넘치는 표현이다. 건축뿐 아니라 거대한 도시 공간에 대한 르 코르뷔지에의 해법도 인간에 대한 탐구와 관심에서 시작되었음이 분명하다. 인간에 대한 그의 존중은 도면 구석에 있는 스케일바에서도 드러난다. 그 스케일바는 거리 단위뿐 아니라 사람의 움직임으로 거리의 정도를 알려주고 있다.

스케일바는 설계하는 입장보다 도면을 읽는 입장에서 더 필요한 정보이다. 하지만 설계자인 그에게도 매우 중요한 것이었고, 그가 설계한 공간들에서 얼마나 세심한 휴먼 스케일의 이야기들이 숨겨져 있을지 예상된다. 지금도 학교와 실무에서 하루도 빠지지 않고 등장하는 현대건축의 거장이지만, 설계된 공간이 주가 아닌 그 공간을 점유할 인간이 주가 되어야 함을 몸소 보여주고 있다.

6

보는 이가 움직이는 건지 그림이 움직이는 건지 M. C. 에셔의 그림들은 신비롭다. 2차원의 평면에 그려졌지만, 표현된 동선들은 평면상의 동선이 아니다. 어디서부터 관찰해야 할지도 모호하다. 혹은 어디에서든 이야기를 시작해도 전혀 상관없다. 돌다 보면 제자리로 와 있다. 뭔가 에셔한테 농락당하는 기분이다. 하지만 기분이 나쁘지도 않다. 그저 그림 한 장을 보며 알 필요도 없는 근거를 찾으려 혼자 애를 쓰다 끝난다.

에셔의 바닥, 벽, 천장은 서커스 저글링같이 빠른 속도로 관계를 바꾸고, XYZ축의 경계를 허물며 중력과 구조를 거부한다. 벽, 바닥, 천장을 구분하고 중력과 구조를 고민하며 설계하는 나와 대조되는 에셔에게서 나는 도전을 받는다.

동선에 큰 관심을 갖고 설계를 시작한 프로젝트가 꽤 있다. 에셔가 상상하는 동선이 내가 구성한 현실의 관점에서는 불가능할지라도 사용자 관점에서는 가능하지 않을까? 동선은 단지 그들이 걸어 다니는 발자국만이 아닌 그들이 공간을 보고 경험하는 계기이기도 하다. 설계자가 구성한 동선과 공간에서 의도되지 않은 상황까지도 사용자가 경험하기를 바라는 마음이다.

김이홍

7

미국 LA에서 인턴 생활을 할 때 한 지인이 방문해 캘리포니아주의 데스밸리(Death Valley) 국립공원에 함께 간 적이 있다. 위치도 정확히 모르는 곳이었고 사막이라 내키지 않았지만, 마지못해 끌려간 여행이었다. 하지만 출발 전 사연과는 달리 지금까지도 뇌리에 깊게 각인된 장소이다.

이 국립공원은 여름철 기온이 전 세계에서 가장 높은 곳(최고 기록 58.3℃) 중 하나이며, 서반구에서 가장 낮은 지점(고도 -86m)에 있는 극단적인 자연환경으로 이루어져 있다. 그중에서도 메스키트 플랫 샌드 듄스(Mesquite Flat Sand Dunes)는 잊을 수 없는 경험이었다. 이튿날 아침 해가 뜰 무렵 바람에 날려 이루어진 사막의 모습은 인간의 힘으로는 만들 수 없는 풍경이었다. 내가 깊게 고민해서 그린 선들과는 차원이 다른 아름다움이었다. 특히 언덕의 능선들이 춤을 추듯 물결치는 모습은 자연법칙에 따른 것이겠지만, 계산할 수 없는 선들의 연속이었다.

나는 어려서부터 수학을 좋아하고 잘했다. 수학으로 대학교에 입학했을 정도였다. 또한 모눈종이를 좋아한다. 꼭 직선만을 고집하는 건 아니지만, 사선이나 곡선을 그릴 때는 합리적인 이유를 찾아야 하고 수학적으로 계산된 좌표에 존재하는 선을 그린다. 그러나 그날 그곳에서는 장엄한 자연 앞에서 규칙을 운운하는 내가 한없이 작게 느껴지는 경험을 했다.

original 911
1963

G-Series
1973

964
1988

993
1993

996
1997

997
2004

991
2011

8

포르쉐 911은 스포츠카의 대명사 포르쉐 라인 중에서도 50년이 넘게 이어오는 핵심 모델이다. 현재까지 7세대를 거친 역사를 지니고 있다. 외관의 디자인뿐 아니라 시대에 부합하는 소비자의 수요 그리고 새로운 기술과 자재의 도입 등 눈에 보이지 않는 부분에서도 변화를 지속해왔다.

무엇보다도 감탄을 자아내는 점은 세대를 거친 디자인의 변화이다. 50년간의 7세대 모델을 부분적으로 콜라주한 포스터를 보면 어색함 없이 완성된 하나의 911모델이 눈에 들어온다. 5-10년마다 바뀌는 세대별 모델에서는 분명 차이가 있었지만, 긴 역사를 놓고 봤을 때는 디자인의 DNA가 바뀌지 않은 유기적인 진화 과정이 보인다.

50년여 간 분명 디자이너 한 명의 손만을 거치지는 않았을 것이다. 이 정도 되었으면 영속적인 가치의 디자인으로 평가되어야 하고, 디자인의 영속성은 변화 과정에서도 굳은 심지가 되어주는 콘셉트에서 기인한다.

시간이 흐를수록 더 완성도 높은 자동차로 진화하고 있는 포르쉐이지만 특이하게도 완성품으로서의 포르쉐 911보다 포르쉐 911모델의 콘셉트가 더 명확해지고 있는 듯하다. 이것이 '자동차'라는 일반명사를 뛰어넘어 '포르쉐 911'이라는 고유명사로 불리는 이유이기도 하다.

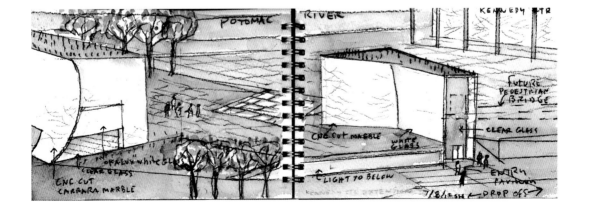

김이홍

9

2013년 1월부터 2016년 5월까지 41개월 동안 스티븐홀아키텍츠에 재직하며 워싱턴DC의 케네디센터(John F. Kennedy Center) 증축 공사에 참여했다. 이때 매일 대면했던 스티븐 홀이 보여준 건축을 대하는 자세는 건축가로서 지금의 나를 형성하는 데 큰 영향을 주었다.

맨해튼의 스카이라인이 보이는 스티븐 홀의 작업실은 매일 그가 스케치하고 글을 쓰며 프로젝트를 구상하고 상상을 펼치는 공간이다. 책상 위의 선반에는 같은 크기의 스케치북들이 순서대로 진열되어 있다. 4×6in의 스케치북은 그가 출퇴근하거나 이동할 때 항상 소지하며 집과 기차, 사무실 어디서든 관찰과 사고를 기록하는 도구이다. 그의 트레이드마크인 수채화 스케치로 투시도에서부터 스케일에 맞춘 상세 도면까지 자세히 담고 있다. 이른 아침 수채화로 하루를 시작하고, 출근하자마자 프로젝트 담당자들에게 스케치를 나눠주는 모습이 지금도 생생하다.

1947년생이고 1977년에 사무실을 열었으니, 그의 스케치북 역사도 40년 이상 되었을 것이다. 그 시간 동안 한결 같은 꾸준함과 열정이 책장에 꽂힌 수십 권의 스케치북에 고스란히 녹아 있다. 이것이 현재의 건축가 스티븐 홀을 있게 한 힘의 원천이라 믿는다.

단
DAN

고즈넉한 신문로2가에서 인왕산을 바라보며
경사진 도로를 따라 걸어 오르다 보면 단(이하 DAN)
프로젝트가 그 단정한 모습을 서서히 드러낸다.
패션브랜드 사옥인 DAN은 첫 미팅부터 전달받았던
이 브랜드의 정체성을 꾸밈없이 그대로 드러낸다.
그리고 결코 가볍지 않은 회사의 의지를 담아
고객과의 소통 그리고 직원 간의 소통에 초점을 두고
설계했다.

육면체에서 시작된 건물의 매스는 북측 사선제한,
도로에서의 진입, 외부로의 조망, 내부 간의 관계
등에 맞춰 조각되기 시작한다. 외부에서 도려낸
볼륨뿐 아니라 내부에서도 매스를 도려내는 방식으로
소통의 장치를 마련했다. 도로와 마주한 입면
3층에 자리 잡은 깊숙이 파인 부분은 외부와 내부의
도려냄이 중첩되는 부분으로 외부-외부, 외부-내부,
내부-외부, 내부-내부로의 다양한 소통을 가능케 하며
연결고리가 되는 중요한 보이드(void)다. 외장재로
마감된 벽의 이면에는 내부 공간과 함께 외부 공간이
등장하기도 한다. 또한 내부에서의 조망이 외벽
개구부를 통해서 간접적으로 이루어지기도 하며
외장재로 마감된 벽체를 바라보기도 한다.

전면 도로에서 I층으로 올라가 원형 기둥들이 나열된
회랑을 거쳐 3층까지 외부 계단을 통해 올라가면
새로운 마당에 도착한다. 회랑에 위치한 정면부의
외부 공간은 외부인에게도 개방되는 반면, 3층 마당은
회사 직원을 위한 건물의 중간 지점에 자리 잡은
안식처이다. 건물로 진입하여 5층까지 자연스럽게
연결되는 외부 동선을 오르다 잠시 쉴 수 있는,
그리고 여러 층에 나뉘어 근무하는 직원이 함께
모일 수 있는 곳이 된다. 바느질하는 실이 건물을
천 삼아 꿰고 올라가는 듯한 동선은 다양한 지점에서
흥미로운 외부와 내부의 풍경을 제공한다. 3층부터
5층은 엘리베이터와 외부 수직 계단을 하나의
코어로 구성해, 내부 업무시설의 공간적 효율성을
극대화했다. 동일한 실들이 적층되는 3-5층은
내부 계단을 통해 직원 간의 협업 및 소통과 이동을
원활하게 돕는다.

2층 바닥 레벨까지 노출콘크리트로 마감된 원형
기둥과 슬래브, 골조 덩어리가 미색 타일로 마감된
건물의 매스를 받치고 있다. 재료에 의한 구분과 함께
규칙적으로 배치된 원형 기둥들로 비대칭적인
상부 매스를 지지하는 모습이 규칙성과 조형성을
구분 짓기도 한다. 인접 건물에 비해 웅장한 존재감의
DAN은 옛 감성이 묻어나는 질감의 타일로 마감되어
신문로2가의 정취가 그대로 느껴진다.

김이홍

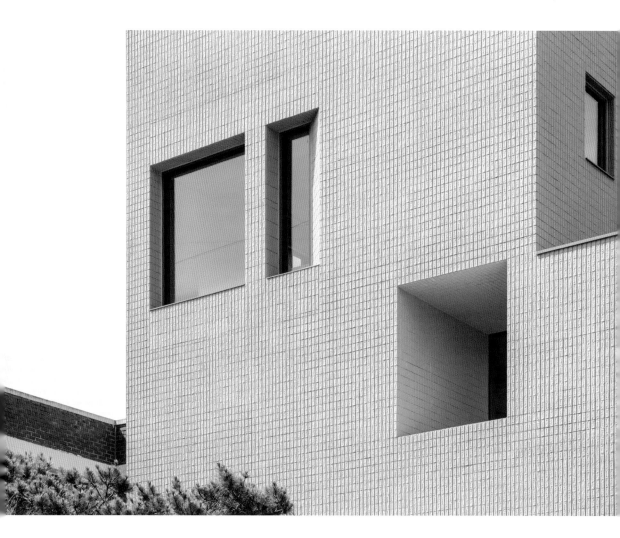

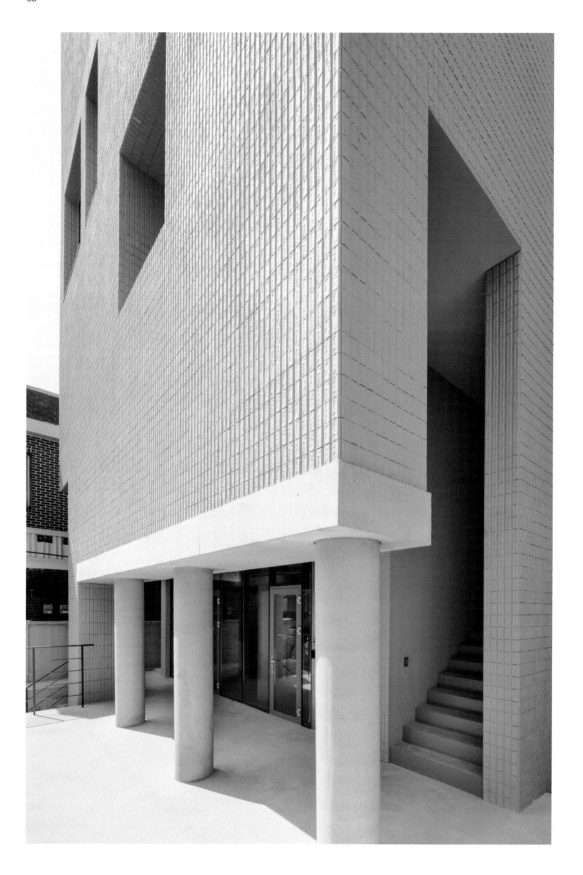

김이홍

김이홍

설계	김이홍
위치	서울시 종로구 신문로2가 1-137
용도	업무시설
대지면적	311.1m²
건축면적	156.69m²
연면적	835.52m²
규모	지상 5층, 지하 1층
높이	16.86m
주차	5대
건폐율	50.37%
용적률	195.97%
구조	철근콘크리트조
외부마감	타일, 스터코플렉스, 노출콘크리트
내부마감	석고보드 위 페인트
구조설계	(주)모아구조기술사사무소
기계설계	지아이이기술사사무소
전기설계	케이에스엔지니어링
시공	(주)이각건설
설계기간	2016. 8 – 2017. 3
시공기간	2017. 6 – 2018. 5
건축주	(주)314호넷
사진	신경섭

평면도
Plan

2F

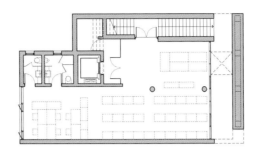

1F

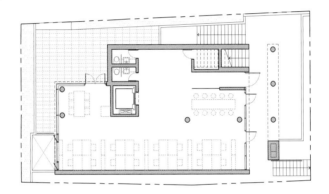

B1

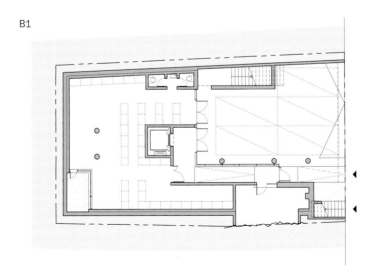

0 1 2   4M

5F

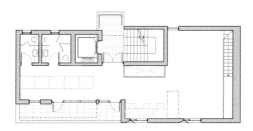

4F

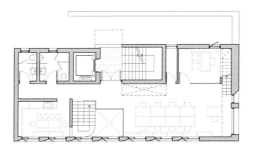

3F

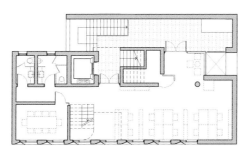

단면도
Section

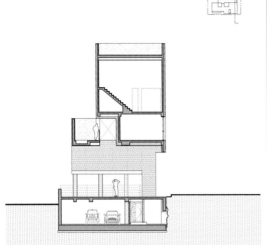

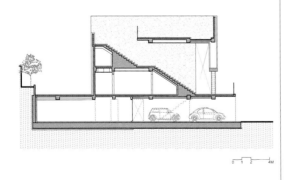

김이홍

입면도
Elevation

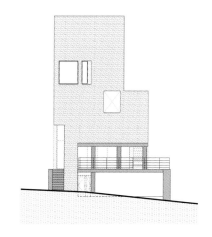

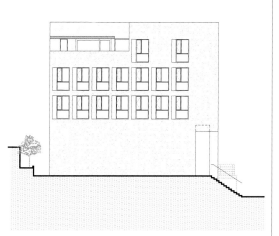

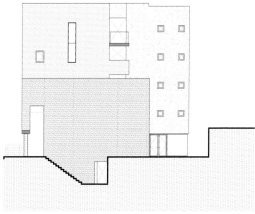

57E130 NY 콘도미니엄
57E130 NY Condominium

57E130 NY 콘도미니엄은 뉴욕의 합벽 건축이 갖는 유형학에 변화를 주어 도시의 콘텍스트를 거스르지 않고 차별화된 디자인을 제안하는 주거 프로젝트이다. 지상 6층, 지하 1층 규모로 총 5개의 유닛으로 이루어졌다.

남북으로 긴 뉴욕시의 맨해튼은 그리드 체계로 구획되어 약 1,300여 개의 블록으로 나뉜다. 이 사이트는 맨해튼 북동쪽의 이스트할렘(East Harlem) 지역에 위치하고, 폭 18ft-3in(5.58m)와 깊이 100ft(30.48m) 규모로 구획된 필지다. 양쪽으로 인접한 건물과 합벽 구조 그리고 연속 파사드로 계획하는 것이 구역상의 기본 조건이다. 결론적으로 뉴욕 신축 프로젝트는 대부분 3차원적 볼륨의 계획보다는 2차원적인 파사드 계획이다. 다행스럽게도 이 사이트는 서쪽과 동쪽에 인접한 건물의 파사드가 나란하지 않고 동쪽 파사드가 2m 가량 후퇴되어 있어 남동측 코너가 입체적으로 드러나는 이점을 갖고 있었다. 제도적 제약과 함께 뉴욕의 파사드 콘텍스트는 57E130 NY 콘도미니엄의 가이드라인을 규정했다. 층별로 4개의 좁은 내력벽으로 3베이 오프닝이 생기는 보편적인 파사드 패턴을 존중하기로 했다. 이 프로젝트는 외장재인 벽돌 낱장의 모듈로 전체 건물 매스 계획과 세부적인 파사드 디테일을 해결했다. 8in(20.32cm) 폭 벽돌 27개로 이 건물 파사드가 구축된다. 그리고 한 층당 40개의 벽돌이 소요된다.

반듯한 육면체 매스가 인접한 건물 사이에 밀착했다기보다는 인접한 건물들에 매스를 퍼즐같이 끼우고 벽돌 디테일로 표현했다. 육면체 매스의 좌측 하단(1, 2층의 좌측)과 우측 상단(6층의 우측)을 벽돌 1.5개 폭만큼 45도 각으로 모따기(chamfering)를 했다.

세부 디테일 측면에서는 파사드에 12in(30.48cm)의 깊이를 주어 평평한 파사드와 차별화를 시도했다. 벽기둥(내력벽)은 24in(60.96cm) 너비이며 창 개구부는 40in(101.6cm) 너비로 구성된다. 벽기둥 너비의 절반은 평평하고 절반은 45도로 꺾여 깊이감이 구현된다. 그리고 45도 각이 생기는 방향은 건물의 전체적인 조화 속에서 방향을 바꾸어 단조로운 3×6 파사드 그리드에 생동감을 불어넣는다. 45도로 꺾이는 부분은 135도의 각을 가진 벽돌이 사용된다. 내부 상황에 맞춰 벽기둥의 너비가 다른 구간이 두 군데 있다. 이 부분은 124도, 146도, 158도의 각을 가진 벽돌이 사용되어 총 다섯 가지의 벽돌 타입으로 전체 파사드가 구성된다. 파사드의 깊이감은 그림자의 연출로 더욱 극대화된다.

김이홍

57E130 NY 콘도미니엄

김이홍

57E130 NY 콘도미니엄

설계	김이홍
위치	미국 뉴욕시 (57 East 130th Street)
용도	주거시설
대지면적	169.36m²
건축면적	107.17m²
연면적	746.71m²
규모	지상 6층, 지하 1층
높이	18.29m
건폐율	63.3%
용적률	343.54%
구조	조적조
외부마감	벽돌, 스터코
내부마감	석고보드 위 페인트
구조설계	A Degree of Freedom
기계설계	New York Engineers
전기설계	New York Engineers
시공	Edgehill Construction, Inc.
설계기간	2014. 3 – 2015. 5
시공기간	2015. 7 – 2018. 8
건축주	Verse Developement
사진	이호경

파사드 다이어그램
Facade Diagram

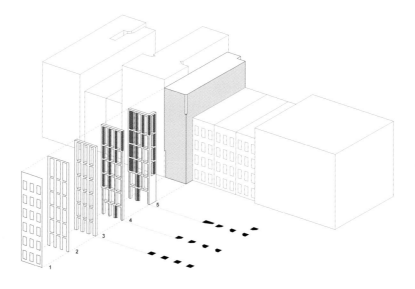

입면도
Elevation

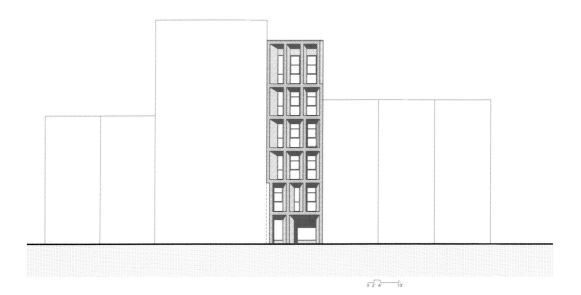

평면도
Plan

6F

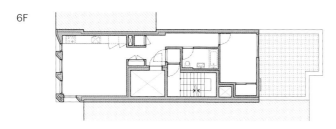

3–5F

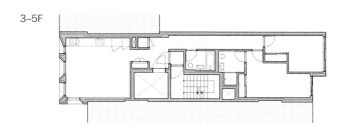

2F

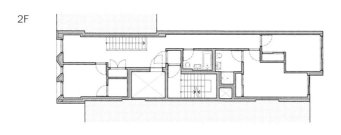

1F

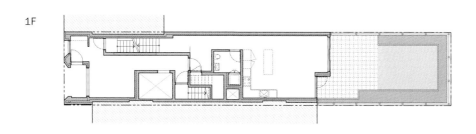

벽돌 타입
Brick Types

| 타입 A | 타입 B | 타입 C | 타입 D | 타입 E |

2F

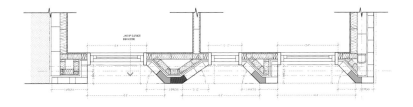

1F

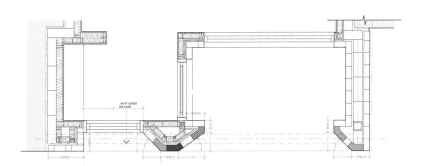

6F

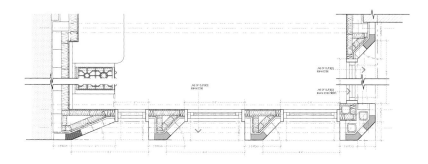

5F

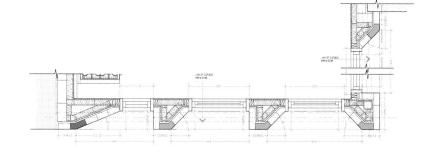

3–4F

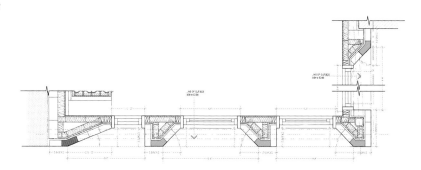

57E130 NY 콘도미니엄

코너스톤 1-532
Cornerstone 1-532

코너스톤 1-532는 북측의 고급 주택지역과 남측 상업지역의 경계에 있다. 우후죽순 생겨나는 다세대주택, 인근의 추계예술대학교와 이화여자대학교 등 북아현동에서 일어나는 발랄하고 무질서한 다양성을 차분히 받아들여 담담하면서도 그 동네만의 매력이 있는 건물이 되길 바랐다.

53평 대지에 건폐율 40%의 넉넉지 않은 신축 조건. 하지만 사거리 모퉁이에 위치한 장점을 활용하였다. 도로와 마주한 연속의 곡면 형태를 택함으로써 건물의 분절을 최소화해 규모의 시각적 극대화를 시도하였다. 또 하나의 전략으로 경사 대지의 고저 차이를 활용해, 1층 레벨을 높여 반지하에 자연채광이 되면서 도로에서의 출입을 가능하게 했다.

1층은 대지경계선을 그대로 본뜬 프레임으로 구성하여 곡면과 대조되는 새로운 지오메트리를 삽입하였다. 프레임은 건물의 매스에만 삽입되는 것이 아니라 도로와 마주하는 대지의 서쪽 면을 따라 캐노피(canopy)로 연장되어 담으로 자연스럽게 이어져, 외관상 건폐율 100%의 효과를 시도하였다.

멀리서 접근할 때는 아스팔트 도로에 둥근 돌 같은 매스가 육중하게 서 있지만, 행인의 시야에서는 수평 프레임에 끼워진 투명한 유리를 통해 건물 내외부가 교감하는 소통의 장치 역할을 한다. 다양함 속에 조화를 이루고자 무채색 전벽돌로 건물의 담담함과 묵직함을, 그리고 노출콘크리트로 프레임의 날렵함과 가벼움을 표현해 질감과 투명도의 대비를 이룬다. 불투명과 투명의 조율은 건축주의 임대 사업성과 건축가로서의 사회적 책임감 사이의 타협이기도 하다.

지하 1층부터 지상 3층까지 건물 형상의 단일 평면이 수직으로 적층되고 지하 1층 후면에 단일실이 추가로 배치되어, 총 5개의 임대 공간으로 나뉜다. 불명확한 임차인을 상상하며 설계했지만, 적어도 반지하 성격의 지하 1층 전면실과 1층 공간은 외부와의 적극적인 연계가 가능하다. 1층 프레임에는 수직 멀리언바를 제거하고 곡면 강화유리로 마감했다. 이 프레임은 코너스톤 1-532 내부의 이야기를 담는 액자이다. 또한 안에서는 대지 형상의 프레임 속에서 곡면 유리 너머로 북아현동의 풍경을 파노라마처럼 즐길 수 있다.

김이홍

김이홍

설계	김이홍
위치	서울시 서대문구 북아현동 1-532
용도	상업시설(카페, 식당, 작가 스튜디오)
대지면적	175m²
건축면적	69.75m²
연면적	310.62m²
규모	지상 3층, 지하 1층
높이	9.7m
주차	2대
건폐율	39.86%
용적률	118.28%
구조	철근콘크리트조
외부마감	전벽돌, 노출콘크리트
내부마감	페인트
구조설계	(주)지우구조기술사사무소
기계설계	선캐리어티앤씨
전기설계	(주)극동문화전기설계
설계기간	2010. 10 – 2011. 2
시공기간	2011. 3 – 2012. 2
사진	신경섭

입면도
Elevation

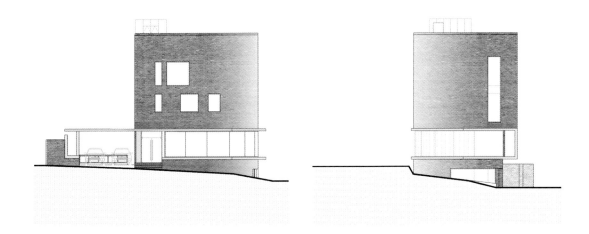

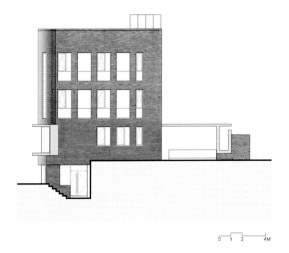

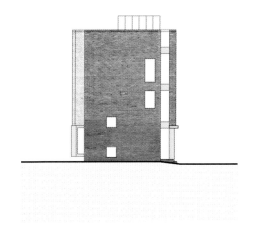

0 1 2   4M

평면도
Plan

2F

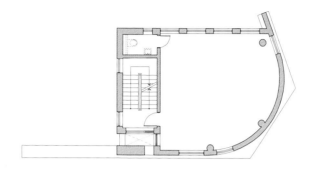

1F

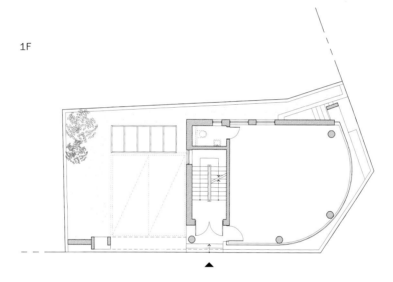

B1

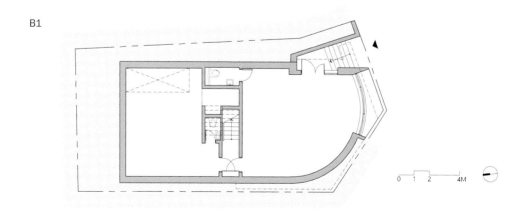

0 1 2   4M

진짜와 가짜 사이: APMAP 2016 용산
Between Real and Fake: APMAP 2016 Yongsan

2016년 여름 아모레퍼시픽 신용산 사옥의 현장은
골조 공사 단계로 한창 분주했고, 현장 부지의 북서쪽
모서리에는 본 건물 시공에 앞서 목업으로 제작된
3층 규모의 건물이 임시로 자리 잡고 있었다.
가짜 건물인 이 목업 건물에 설치된 '진짜와 가짜 사이'
설치물은 건물에서 일어나는 일반인의 당연한 인식에
엉뚱한 전환점을 제공한다.

우리는 건물에서 너무 많은 것을 익숙하고 당연하게
여기며 생활하는데 이 작품은 예측을 뒤집는 반전으로
새로운 경험을 선사한다. 작품이 설치된 구조물은
건물의 형태를 갖추고 있지만 실제 기능하는 건물은
아니다. 문, 창, 문고리, 스위치, 조명기구, CCTV 등의
요소들이 존재하지만 작동하지 않으며 모양새만
갖추고 있다. 진짜 같은 가짜 건물에 겉모습과
이질적인 내부 모습을 제안함으로써 가짜 건물임을
고발한다.

입구에 들어서자마자 두 개의 엘리베이터 문을
마주한다. 첫 번째 입구로 발걸음을 유도할 엘리베이터
버튼을 누르면서 작품이 시작된다. 문을 열면 편안한
엘리베이터가 기다리는 게 아니라 더운 날 걸어 올라야
하는 가파른 계단이 첫걸음을 재촉한다. 진입구로
유도되어 밖에서만 열릴 수 있는 문으로 들어서는
순간 여정을 시작할 수밖에 없다. 그리고 동선을 따라
두 번째 엘리베이터 입구로 다시 나오게 된다.
짧지만 순간적으로 감정의 변화가 흥미롭게 전개된다.
호기심에 들어섰다 좌절감을 갖게 되고, 다시 외부로
나와 안도한다. 좌절감이 절정에 치닫는 계단참에서
가느다란 개구부를 통해 진짜 건물을 전망하게 되고,
진짜 건물과 가짜 건물의 차이를 몸소 깨닫는다.

주재료로 철재와 주황색 PVC 메시(mesh)가
사용되었다. 주황색 PVC 메시는 인접한 진짜 건물
현장에서 작동하는 주황색 공사용 엘리베이터와
연계한다.

설계	김이홍
위치	서울시 용산구 한강대로 100
용도	공공미술
규모	405×420×500cm
높이	5m
구조	철골조
외부마감	철판, 철파이프, PVC메시
시공	테림, 종로천막기업
설계기간	2016. 6 – 7
시공기간	2016. 8
건축주	아모레퍼시픽미술관
사진	배지훈

단면도
Section

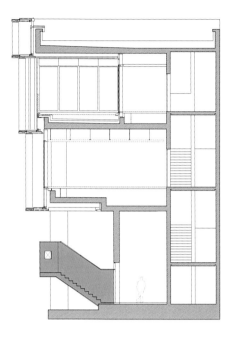

평면도
Plan

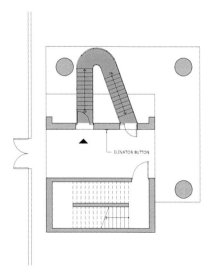

개념작업: 숨바꼭질, 미로
Conceptual Work: Hide and Seek, Labyrinth

숨바꼭질

가상으로 설정한 두 주인공의 심리 관계를 공간화한 '개념 작업'이다. 두 주인공은 서로가 상대방이 되기를 갈망하는 심리적 갈등을 겪는다. 각각의 심리를 단일 상자에 비유하고, 순환되는 심리를 상자 안에 다른 상자가 있는 방식으로 표현해 내부에 무한하게 반복되는 상자로 구체화한다. 3개의 다른 축에 있는 면들이 모이면 최소한의 공간이 만들어지는 구축 원리처럼 3개의 유닛을 XYZ 축에서 끼워 맞춰 공간을 형성한다. 끼워 맞춰질 3개의 유닛은 12×12in 규격의 배스우드 판재에서 시작한다. 각 유닛은 판재 ¼만큼을 90도 각으로 접은 뒤 접힌 면 ¼만큼을 다시 접는 반복되는 과정을 거쳐 만들어진다. 동일한 3개의 유닛을 끼워 맞추면 연속적으로 내재되는 상자의 형태가 만들어진다. 무한하게 작아지는 상자를 통해 모든 인간의 헛되고 허무한 내면을 보여주기도 한다.

미로

피카소의 1935년작 '미노타우로마키(Minotauro-machy)'에 등장하는 미노타우로스(Minotauros)와 테세우스(Theseus)의 관계를 해석해 공간화한 '개념 작업'이다. 고대 그리스 신화 속 영웅인 테세우스가 반인반수 괴물인 미노타우로스를 퇴치하러 그가 사는 미로 공간으로 들어가 일방적으로 쫓는 관계로 그려진다. 하지만 나는 미노타우로마키에서 멈춰 있는 두 주인공이 서로를 향해 뻗고 있는 팔에 의해 양방향으로 쫓는 관계로 해석했다. 이 미로는 결코 만남을 갖지 못하는 공간이며, 두 주인공의 동선을 2개의 고리가 얽혀 있는 형상으로 구체화한다. 동일한 공간에 있는 것이 아닌 각자의 공간에 갇혀 있는 두 주인공이다.

투명한 유리벽을 통해 서로의 존재를 인지하며 뛰어다니지만, 얽힌 2개의 고리처럼 접점이 없으며 물리적으로는 만날 수 없는 공간이다. 공간 설계는 25×2.5in 규격 월넛 원목판재 10개와 그와 동일한 크기의 아크릴판 1개로 시작한다. 이는 월넛 원목판재를 적층해 만들어진 매스의 일부분을 잘라내고 이동하며 공간을 만들어가는 방식으로 발전된다.

긴 축으로 평행한 3개의 통로가 만들어지고 아크릴판은 통로 사이의 벽체로 삽입되며 부분적으로 투명한 면을 통해 시각적 소통이 이루어지는 접점이 된다. 이 공간에 들어간 미노타우로스와 테세우스는 다시 나오지도 못하고 서로를 잡을 수도 없는 미로에 갇힌다.

투시도(숨바꼭질)
Perspective (Hide and Seek)

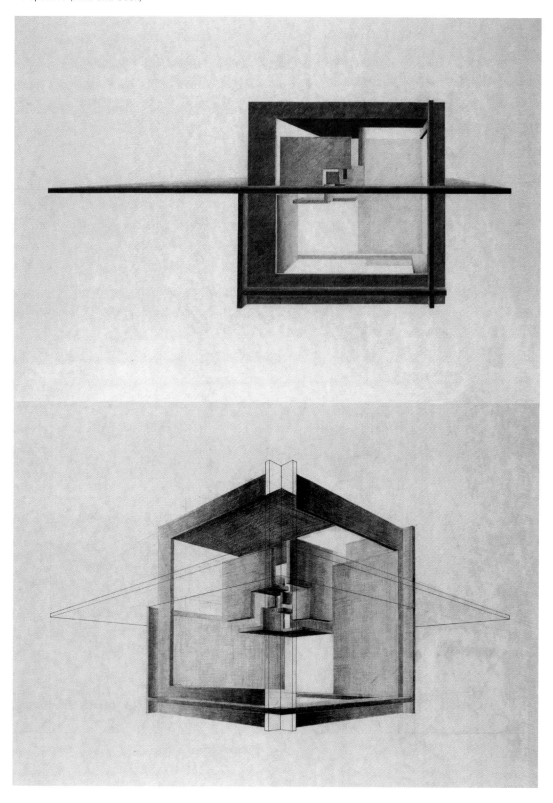

프로세스 다이어그램(숨바꼭질)
Process Diagram (Hide and Seek)

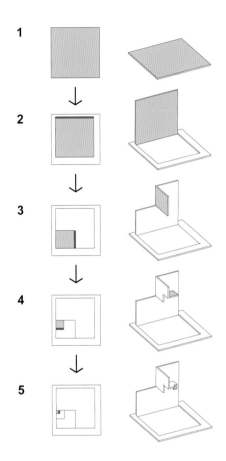

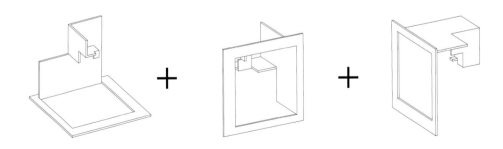

모형(숨바꼭질)
Model (Hide and Seek)

스케치(미로)
Sketch(Labyrinth)

프로세스 다이어그램(미로)
Process Diagram (Labyrinth)

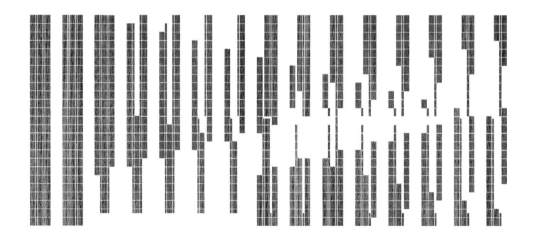

모형(미로)
Model (Labyrinth)

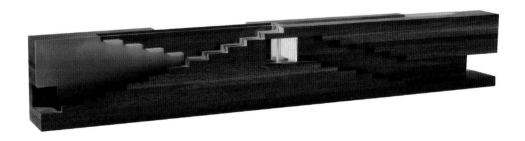

비평

허구와 실재 사이
—김이홍의 건축
그리고 현대의 딜레마에 대하여

존 홍 · 서울대학교 교수

'허구와 실재 사이'가 이 글을 위해서 만들어진 제목이라고 주장할 수
있다면 좋겠다. 단순하고 거친 표현이지만, 이 제목은 오늘날의 젊은
건축가가 처한 현대의 딜레마를 잘 함축하고 있기 때문이다. 그들은
프로젝트의 개념적인 기원에 진정성을 담아내면서 프로젝트가 진행되는
동안 점점 복잡해지는 예기치 못한 사태들과 계속 협상해야하는 상황에
놓여있다. '허구와 실재 사이'라는 제목은 건축가 김이홍이 2016년에
아모레퍼시픽 본사의 공사 현장에서 진행했던 전시 작품명이며 불확실한
상태를 의미한다.신진 예술가를 위한 전시로, 데이비드 치퍼필드(David
Chipperfield)의 건물이 지어지는 동안 현장에 설치되었다.

제목이 전조하고 있듯, 김이홍의 작업은 관객에게 자칫 따분하게
느껴지는 상황 속에서 일종의 환희를 선사한다. 관객은 맨 처음
2개의 엘리베이터 출입구를 마주한다. 그리고 이것이 매우 일상처럼
보이면서도 기대감과 불안감을 함께 느끼게 하는, 짐짓 어딘가 불편한
장면임을 인지한다. 전시 관람은 엘리베이터 버튼 같은 장치를 누르는
행위로, 그들이 지금까지 일상 속에서 수없이 마주했을 순간의 반복으로
시작한다. LED 지시등에 불이 들어오면 관객은 '리프트' 안으로 걸어
들어간다. 하지만 그곳에서 그들을 기다리고 있는 건 주황색 공사용
가림천으로 쌓인 초현실적 계단이 전부다. 그리고 그 계단의 디자인은
여전히 공사 중인 아모레퍼시픽 건물을 지표로 삼아 변주된다. 우리가
보통 사람의 접근을 막기 위해 공사 부지 주변을 둘러싸는 일반적인
재료가 이곳에서는 사람이 들어가야 하는 통로 공간을 감싸며 완전히
반대 방식으로 사용되었다. 이와 같이 잘못된 방식으로 사용된 가림천은
햇빛이 투과되면서 외부의 풍경 위에 무아레 무늬를 겹쳐놓고, 관객이
느끼는 일상의 감각과 생경함을 은연중에 더욱 뒤섞어놓는다.
빛이 들어오는 개구부를 향해 이 계단을 따라 올라가다 보면,

아모레퍼시픽 본사 공사 현장의 단편들을 보여주는 좁은 창들이 있는
작은 계단참에 다다른다. 보통의 계단참과는 달리 이 공간은 잠시 숨을
돌리기 위한 장소이기보다 하나의 연장된 통로에 가깝고, 관객은 마치
계속 앞으로 나아가야만 할 것 같은 느낌을 받는다. 또한 이 굽어 있는
공간은 방향감각에 혼란을 주어 관객은 끝까지 가고 나서야 그들이
도착한 곳이 결국 맨 처음 출발했던 곳, 가짜 엘리베이터 로비에 있던
바로 옆의 엘리베이터 문이라는 사실을 깨닫는다.

제목과 현실의 결과물, 두 측면에서 모두 '허구와 실재 사이'는 김이홍의
다른 작업들을 관통하는 강력한 은유이다. 역설적인 상황을 이용함으로써
실제의 물리적 작업물은 가상의 프로젝트이자, 어떻게 건축을 규정할
것인가에 대한 질문의 핵심을 찌르는 하나의 이론적인 선언이 된다.
그리고 이는 건축이 그 유구한 역사 속에서 계속하여 현실과 개념 사이를
넘나들어온 방식과 접목된다. 원본의 끊임없는 복제를 통해 만들어지는
허구에 대한 현대적 비평이 장 보드리야르(Jean Baudrillard)의
『시뮬라크르와 시뮬라시옹(Simulacra & Simulation)』(1981)과 같은
책들에 의해 주도되기 시작한 순간부터, 창조와 재창조에 대한 논의는
항상 건축의 진정성 문제와 결부되어왔다. 건축은 무언가를 관념적으로
상징하거나 동시에 물리적으로 표상함으로써 언어적 특성을 갖게 된다.
그리고 이는 신이 머무르던 형이상학적인 장소들에 존재하는 고대
종교적 구조물에까지 그 기원을 거슬러 올라간다.

긴 역사에 걸쳐 허구라는 개념이 수없이 반복되어왔다는 것을 우리는
종종 잊어버리곤 한다. 어쩌면 건축이 점점 그 영향력을 잃어가는
오늘날이 이 개념을 다시 중시해야 할 시점일지도 모른다. 고대 그리스
건축가들조차 허구의 개념에서 자유롭지 못했다. 그들은 투시도법상의
눈속임을 위하여 엔타시스(entasis)라 알려진 기둥 중간을 살짝 굵게
만드는 기법을 사용했다. 이는 그들의 물리적 구조가 온전히 순수하다고
주장할 수 없게 만든다. 하지만 오늘날과 같이 작가가 의식적으로 특정한
효과를 위해 인공물을 배치하는 모습은 15세기 르네상스 시대에 시작되어
18세기 '네오' 운동들(신고전양식, 신고딕양식, 신바로크양식처럼)과 함께
발전된 허구의 방식에서 시작되었다.

20세기 이후, 잃어버린 진정성에 대하여 죄책감을 가지던
모더니즘의 시대를 뛰어넘어 건축을 통한 재현은 패스티시(pastiche),
캐리커처, 그리고 콜라주 등에 관한 이론과 함께 다시 시작되었다.
포스트모더니즘의 영향이 이런 상황을 그 이전보다 호전했는지 혹은
악화했는지는 불명확하다. 하지만 포스트모더니즘이 던지는 질문은
여전히 본질적이며 아직 그에 대한 논의는 끝나지 않은 상태로 남아

있다. 이것이 김이홍이 '시뮬라크르'나 '아티피스(artifice)'와 같은
보다 간접적인 단어들 대신에 '허구'라는 단어를 선택한 이유이자 곧
신진 건축가로서의 그를 이해하는 핵심 지점이다. 이는 새로운 세대의
디자인의 잠재적 동력이 어떻게 형성될지에 대한 통찰을 제시하기
때문이다. 허구는 더 이상 학계에서 벌어지는 둔감한 논쟁들 뒤로 숨지
않는다. 허구는 단지 있는 그대로 허구 그 자체일 뿐이다.

이러한 경향은 지적 논의 속에서 허구성을 공허한 것으로 비워 내버리지
않고, 오히려 허구성을 더욱 자유로운 지위로 승격한다. 허구성은 의식
차원에서 구축되며, 보는 이들을 기만하기보다 그 속임수의 일부가
되도록 그들을 불러들인다. 그리고 여러 사람이 집합적이고 다양한
방식으로 참여할 수 있는 하나의 공간적 놀이터가 된다.

이와 같은 방식으로 김이홍의 작업은 보드리야르가 주창한 위기의
연속선상에 위치한다기보다는 팔라디오, 더 정확히는 팔라디오(Andrea
Palladio)의 제자 빈첸초 스카모치(Vincenzo Scamozzi)가 1585년에
완성한 올림피코 극장(Teatro Olimpico)과 궤를 같이한다고 볼 수 있다.
팔라디오는 오래된 요새의 윤곽 안에 자리한 좁은 대지에 대응하기 위해
로마 원형 극장이 가져왔던 순수한 기하학적 형태를 왜곡했다. 그는
이를 통해 극장을 대지 안에 끼워 넣었을 뿐만 아니라 동시에 배우와
관객 사이에 새로운 거리 관계를 만들어냈다. 그리고 이후 스카모치는
팔라디오가 아레나의 길이를 축소했듯이 무대를 축소해 짓눌린 투시를
만들어냈고, 결과적으로 팔라디오의 이론과 프로젝트를 완성하게
되었다. 관객이 봤을 때 이 왜곡은 명백하게 인위적이지만 그럼에도
강력한 깊이감을 만들어낸다. 또한 원근법적으로 극장의 실내를
외부의 도시 거리와 연결한다. 마찬가지로 김이홍의 작업에서도 작가가
이러한 시각적 '속임수'를 사용해 자신의 의도대로 도시의 풍경과 관객
사이에 접점을 만들어내고 있는 것을 발견할 수 있다. 이를 통한 추상적
개념들은 현실 세계의 제약들을 가로지르며, 방문자는 여기에 유쾌하게
맞물릴 수 있게 된다.

유형과 무형 사이: 코너스톤 I-532

김이홍의 코너스톤 I-532 프로젝트로 예를 들면, 이 디자인은 허구와
실재의 사이, 추상과 물질의 사이를 몇 가지 주요한 방식으로 계속
넘나들고 있다. 175m²의 작은 대지에 자리하는 이 건물은 법적으로
서울의 상당 면적을 차지하는 근린생활시설로 분류된다. 이 분류에 속한
건물 대부분은 건축가의 특별한 디자인 없이 지어진다. 이는 주어진 건축
밀도를 채우기 위한 보편적 형태의 상자들이기 때문이다. 김이홍은 대지

조건을 프로젝트의 제약 요소로 단정 짓지 않고, 반대로 코너 부분의 대지 경계선을 현명하게 활용한다. 땅을 '최대한으로' 채우기 위해 제멋대로 깎여진 경계선으로 건물의 형태를 만들지 않고, 의도적으로 건물의 덩어리를 부드러운 곡면으로 처리했다. 이 모호한 크기의 곡면은 보는 이가 그 끝을 가늠할 수 없게 만든다. 대지 경계선의 깎인 모서리를 건물의 형태로 그대로 반복했다면 이는 오직 대지의 물리적 경계를 다시 강조하는 일밖에 되지 않았을 것이다. 여기서는 이 곡면이 본래의 대지 경계선을 감추어버린다.

눈속임은 여기서 멈추지 않는다. 곡면은 살짝 연장되어 지지물 없이 홀로 서 있는 서측 벽까지 이어진다. 이 측벽은 인접 부지로부터 반드시 후퇴해야 하는 곳인데, 곡면만이 이곳까지 연장되면서 건물 전체 덩어리를 덮는다. 그리고 이 벽이 대지 모서리와 만나면서 그 뒤 건물 서측 입면을 가리게 되어 결과적으로 건물이 실제보다 더 크게 보인다. 이러한 숨바꼭질 놀이의 일환으로, 곡면은 큰 창 하나 없이 벽돌의 픽실레이션만으로 인식된다. 가파르게 경사진 주변 길에서는 건물의 높이를 거의 파악할 수 없고 때로는 정체를 알 수 없는 거대한 돌기둥처럼 보인다.

건물은 전체적으로 무겁고 드럼통 같은 육중한 면모를 갖지만 1층의 경우 놀라울 정도의 수평성을 보인다. 상부의 벽돌이 주는 강한 무게감에 대항하듯, 안의 빈 공간은 세심하게 곡면 처리된 투명한 유리로 감싸져 있다. 이 두 재료는 변증법적 관계로 서로 공명하며, 동시에 다른 한쪽을 부정하고 압박한다. 일종의 장난과 같이 수직 실리콘 조인트와 상하부 비노출형 멀리언(mullion)으로 지지되고 있는 이 비물질적인 유리는 그 위의 물질적이고 공예적인 벽돌들을 반대한다. 유리는 건물의 텍토닉적 성격을 부정하고, 벽돌은 이를 강화하는 것이다. 그리고 그 둘은 함께 실재와 허구의 개념을 표상한다. 그것은 마치 1층의 유리로 둘러싸인 빈 공간이 벽돌 곡면이 만들어내는 시각적 놀이를 물질적으로 그리고 동시에 비유적으로 드러내고 있는 것과 같다. 마지막으로, 콘크리트 인방이 간결하게 지나가면서 벽돌과 유리를 직접 만나지 않게 하며, 실재와 허구 사이의 공간을 표현한다. 이 얇은 선은 벽돌과 유리면처럼 휘어 있지 않고 깎아진 대지 경계선 모양을 순순히 따르고 있다. 이는 비현실적인 순간들 사이에 대지 조건이라는 현실의 치수를 끼워 놓은 것 같이 보인다.

두꺼움과 얇음 사이: DAN

김이홍은 이 주제를 그가 자신의 건물을 디자인하기 이전부터 중요하게 생각했다. 그는 여러 다른 사무소에서 일하는 동안 모델을 통해 공간에

대한 자체 실험들을 진행해왔다. 그 모델들은 심지어 '건축물'이라 불릴
수 있을 정도로 크고 매우 세심하게 만들어졌다. 이는 드로잉이 갖는
덧없음과 건물이 갖는 구축적인 확실함의 사이에서 중간의 입장을
취한다. 그의 동료들이 현상공모라는 '큰 성공'이 걸린 도박에 투자하고
있을 동안, 김이홍은 조용하게 그의 개념적인 능력을 기르고 다듬고
있었다. 마치 미래에 다가올 사건들에 대비하고 있던 것처럼 말이다.

예를 들어 2011년부터 시작된 그의 '숨바꼭질' 연구는 그가 한국의 대형
설계사무소인 삼우종합건축사사무소에서 근무할 당시 진행되었는데,
2018년에 준공된 패션브랜드 사옥 DAN의 선행 연구라고 볼 수 있다.
얇은 판들이 만들어내는 기발한 효과를 탐구한 이 모델은 4개의 얇은
판을 접어서 육면체 볼륨을 만들고 이들을 덧붙여 가는 방식을 통해
마치 덩어리를 파내 만든 것 같은 형태를 구축한다. 이 연구는 정해진
프로그램 없이 오직 건축적 의미만을 함축하고 있다. 이는 어떤 문제에
대한 특수해라기보다는 미래의 여러 상황 속에서 다양한 방식으로 응용이
가능한, 일종의 답이 내려지지 않은 공간 연구에 가깝다. 가장 중요한
사실은 이 연구는 완결되지 않았기 때문에 실제 작업들에서 반복되면서
더욱 정교히 다듬어질 가능성을 내포하고 있다는 점이다.

젊은건축가상의 존재 의의 중 하나는 바로 열망을 성공적으로 실현하는
작가를 발견하기 위함이다. 2018년, 김이홍은 7년이라는 짧은 시간
뒤 DAN 건물에서 '숨바꼭질'의 첫 번째 버전을 실현한다. 그는 면을
접어나가는 방식을 똑같이 사용해 프로젝트의 평면과 단면을 일련의
파내어진 빈 공간들로 구성하였다. 이 빈 공간들은 이용자의 몸의 위치와
시야에 맞게 정확히 위치해 있으며 가상으로만 존재하는 전체적인
매스를 관통한다.

파사드에서부터 논의를 시작하자면, 주차 공간에서 반 층 올라가는
계단이 건물 좌측의 두 층 높이의 보이드를 파내고 있는 것처럼 보인다.
여기서 시작된 보이드는 건물의 앞면을 가로지르는데, 이 사옥의 주
출입구를 드러내는 주랑으로 그 영역이 정해진다. 이 보이드는 이어서
건물의 우측면을 파고 올라가 위층까지 연결된다. 방문객을 건물
한쪽부터 반대쪽 끝까지 움직이게 만드는 '비효율적' 동선은 물론 대지의
복잡한 지형, 법규 제약, 그리고 건물의 후퇴선이 만들어낸 결과물이다.
그럼에도 이런 현실적인 제약들은 짐짓 장난스럽게 실제 건물로 풀어지며
방문객을 개념적인 속임수 속으로 끌어들인다.

주목할 만한 점은 3층의 수수께끼 같은 깊은 보이드가 방문객을 일종의
'숨바꼭질' 놀이로 이끌면서 보이드가 형성된 숨은 논리를 발견할 수

있도록 유도한다는 점이다. 마침내 그곳에 도달했을 때 방문객은 그들의
노고를 두 번 보상받는다. 첫째로는 이 보이드가 3층에서도 파사드 수평
방향으로 파내어져 있다는 것을 발견하게 되고, 둘째로는 이 보이드가
1층의 주랑으로 다시 연결된다는 것을 깨닫기 때문이다. 이 보이드는
일종의 개념상의 웜홀인 셈인데 입구와 상부층을 시각적으로 바로
연결하지만, 방문객의 신체는 그곳에 다다르기 위하여 여전히 건물
전체를 횡단해야 하기 때문이다. 평면에서 이 보이드는 그다지 중요하지
않을 수 있으나, 이는 보는 이들이 비워내는 언어를 통해 만들어진
건물의 두꺼움이 사실은 얇은 면들을 접어서 만든 것임을 깨닫게 하는
열쇠이다. 김이홍이 정확히 '숨바꼭질' 모델을 통해 구현한 것처럼 말이다.

코너스톤 1-532에서와 같이 이 보이드 옆 파사드 일부는 홀로 서 있는
벽이 되어 느슨하게 외부 공간을 규정하고 외벽의 두꺼움과 얇음을 더욱
부각한다. 파사드의 물성은 이 허구성을 더욱 강화한다. 서울 도처에서
볼 수 있으면서 어디에서나 사용할 수 있는 건축용 타일을 반절로 잘라
사용했으며, 이 새로운 치수의 타일은 기존의 보편적인 치수를 환기한다.
이 타일은 3차원상으로 보이드를 뒤덮고 있으며, 일상적인 타일에 대한
우리의 보편적인 관념에 의문을 던지게 한다.

3층에서 시작하는 계단은 4-5층으로 연결되는데, 건물 후퇴선 때문에
얇아진 건물의 몸체를 뚫고 새로운 보이드를 만들어낸다. 부동산
가격이 높은 지역의 작은 필지임에도 김이홍은 3층과 5층까지 이어지는
천장에 몇 개의 작은 개구부를 가까스로 집어넣었고, 이는 천창이 되어
업무 공간에 자연광을 쏟아 넣는다. 이 보이드들은 비록 작지만 공간을
극적으로 만들며 좁게나마 외부로 열리는 시야를 확보한다. 결과적으로
이 작은 건물이 실제보다 커보이는 착시를 불러일으킨다.

디자인 논리를 일관성 있게 보여주는 이 세심한 여정을 따라가다가
5층에 다다르면 보이드와 면 사이의 공간 놀이가 희미해지는, 작가가
의도한 극적인 순간을 맞이한다. 덩어리를 비워내는 전략을 철저하게
따르던 건물에, 계단참에 한 사람이 서 있을 정도의 캔틸레버가
더해지는 유일의 순간을 마주하게 되는 것이다. 본능적으로 이끌리는
이 장면을 통해 우리는 도시의 풍광을 바라볼 수 있을 뿐만 아니라 동시에
보이드가 건물을 관통하는 일련의 과정을 다시 뒤돌아볼 수 있다. 이는
마치 건축가가 지금까지의 경험이 모두 하나의 게임이었다는 것을
고백하는 것과 같다. 우리가 이 눈속임에서 잠시 동안 한 발짝 벗어나서
프로젝트의 공간이 어떻게 작동되고 있는지를 면밀히 살펴보고 더
나아가 분석할 수 있도록 한다.

건물에서 나와 도시의 바닥으로 돌아온 뒤 다시 건물을 바라보면, 1층의
주랑이 1980년대 포스트모더니즘 논의로 자칫 빠질 수 있는 위험한
게임을 하고 있다는 것을 알 수 있다. 하부에 지지 구조 없이 서 있는 가짜
콘크리트 기둥은 양 옆의 2개의 진짜 기둥과 정확히 똑같은 텍토닉으로
만들어져 있다. 주랑이 일반적으로 만들어내는 특유의 리듬을 위해 그
사이에 끼워 넣어져 있다. 겉보기엔 대수롭지 않아 보이지만 이는 사실
건축가의 숙련된 솜씨를 반증하는 장면이다. 이는 건축이 구축적 열망의
결과가 아닌, 가장 중요한 언어적 놀이라는 것에 대한 명확한 선언이다.
그러나 피터 아이젠만(Peter Eisenman)의 웩스너 시각예술센터(Wexner
Center for the Visual Arts) 내부의 계단 위에 불가능한 형태로 떠 있는,
강렬하게 시선을 붙잡는 유명한(오명이 난) 가짜 기둥과는 다르게,
김이홍의 기둥은 일반적이고 흔한 유형학적 배경 속으로 녹아들어 유심히
살펴보지 않는 한 알아차리기 어렵다. 이러한 측면에서 김이홍의 언어는
분열보다는 어떤 변화를 부드럽게 재촉하는 것에 가깝다. 그가 만들어낸
게임을 거부하기 전까지 우리는 계속해서 거기에 이끌리고 빠져들게 된다.

미로와 일상 사이: 57E130 NY 콘도미니엄 ─────────

김이홍의 최근작은 뉴욕 할렘에 위치한 매우 얇은 형태의
콘도미니엄이다. 다른 작업들과 마찬가지로 이 건물은 그가 스스로
진행한 가상의 연구 중 하나인 '미로'(2010)와 긴밀하게 연결되어 있다.
이 초기 작업에서 그는 한 종류의 모듈을 단순한 방식으로 쌓아 일련의
복잡한 공간을 만들어내려 했다. 이 공간들은 서로 얽히고 관통하지만
직접 닿지는 않는다. 작은 스케일의 나무 블록으로 만들어진 이 '미로'
모형은 그의 다른 연구들처럼 직관적인 재료의 입체 모형으로, 공간을
실험해보기 위한 개념 작업이다. 이 연구의 결과를 더욱 발전시키기
위해 그는 실제 대지가 없음에도 그 물체의 형태를 길게 제약하는 경계를
가상으로 만들어냈다.

이 연장된 '미로'의 형태는 김이홍이 8년 뒤 설계하고 완공하게 되는
뉴욕 타운하우스와 강한 유사성을 지니고 있다. 뉴욕의 지역제와 건축
법규는 전 세계에서 가장 엄격하기로 악명 높다. 더구나 대지는 건물들이
합벽을 공유해야 하는 제로 로트 구역(zero-lot line)이었기에 제약은 더욱
심해졌다. 이러한 상황에서는 시공의 난이도도 올라가고 채광과 환기에
관한 법적 요구 조건을 만족하기 위한 해결책 역시 매우 복잡해진다.
그의 '미로' 작업 때처럼, 상황을 어렵게 만든 건 폭이 겨우 5.5m밖에 되지
않는 길고 좁은 대지 안에 독립된 개별 입구를 가지는 5개의 주거 유닛을
짜 맞춰 넣어야 하는 조건이었다. 또한 이용자의 접근성을 중요시하는
뉴욕의 승강기 법규에 맞게 큰 규모의 엘리베이터가 필요했고, 보다 큰

규모의 건물과 동일한 규모와 수준의 비상계단을 요구하는 법규가 이 복잡한 상황의 대미를 장식했다. 이는 논리적인 평면 계획을 불가능하게 만드는 장애물들이었고 난해한 요구 조건들 자체가 미로인 셈이었다.

평면상 공간의 복잡성을 드러내기 위해 '미로'에서 여러 레이어를 자른 것처럼, 할렘의 타운하우스에서 역시 프로그램과 동선이 퍼즐처럼 짜여진다. 부동산 개발업자 주도로 진행되는 이 같은 건물 유형은 보통 법규를 최소한으로 만족하는 데 집중하느라 건물 디자인 자체는 어쩔 수 없이 무디게 마무리하는 경우가 많다. 하지만 김이홍은 이전 이론 작업의 벽돌 모듈이 서로 맞춰 끼워지는 것과 같은 방식의 공간적 논리를 이곳에 적용하고 있다. 방들의 모서리가 사라지며, 방이 아니라 작고 철저히 픽셀처럼 쪼개진 공간들이 창고, 채광, 입구, 비상계단을 만들어내고 각 공간의 질을 향상한다. 다른 건축가라면 명확하고 다이어그램적 공간 조직을 추구할지도 모르는 지점에서, 김이홍은 미로와 같은 복잡함을 만들어내고 이를 수용한다. 다시 말해 그는 '효율성'이라는 기준으로 해결책의 가짓수를 좁히지 않고, 평면의 복잡성을 공간 생성 개념으로 활용해 실내 공간의 전체적인 질을 향상할 수 있는 해결책을 고안했다. 이 지점에서 현실 속의 건축 법규들과 초현실적인 미로 속 공간들은 하나로 합쳐진다.

벽돌을 다시 살펴보면, 김이홍은 이 재료를 기본에 충실하면서도 영리한 방법으로 파사드에 적용했다. 그의 건물의 개구부들은 주변의 인접한 도시 경관에서 쉽게 발견할 수 있는 반복적이고 자기 참조적인 형태의 창문들이지만, 그는 모서리가 깎인 벽돌 모듈들을 조합해 이들을 두껍게 만들었다. 그리고 이 경사지게 깎인 모서리들이 개구부의 한 측면 또는 반대쪽 측면으로 미묘하게 달라지면서 창문의 격자 구조를 만들어낸다. 이 접힌 면은 튼튼한 벽돌을 때때로 종이처럼 얇게 보이게 만들며 입면 전체에 깊이감을 더한다. 이전의 프로젝트들과 마찬가지로 이 작업 역시 디자인의 변화가 극적으로 드러나지 않으며, 따라서 보는 이에게 분석적으로 인식되기보다는 하나의 인상으로 다가간다. 여기서 허구와 실재 사이의 경계는 그 차이를 크게 드러내기보다 앞뒤로 복잡하게 얽혀 있다. 우리는 처음에 파사드 디자인이 가진 균형 잡힌 규칙성(어떻게 보면 보편성)을 마주하게 된다. 격자 구조가 완전히 정렬되지 않았다는 사실에 대해 누군가 굳이 물어보기 전까지 이 파사드에 대한 인상은 아마 지속될 것이다. 계속 보다 보면, 파사드가 가진 무거운 물성은 순간의 놀이를 위한 얇은 포장지로 변한다. 만약 건물이 처음부터 가벼운 재료로 만들어졌다면 이 불안정함은 이처럼 특별하지 않았을 것이다. 흥미로운 사실을 하나 더하자면, 시공에 들어가기 전 김이홍은 대지 안에 차례로 솟아오르는 듯한 빨간 풍선들로 구성한 설치 작품을 짧은 기간

동안 전시했다. '할렘 축하하기'라는 이름의 이 작업은 원래 그 자리에
있던 다 허물어져가는 건물에 대한 일종의 추모식이었고 동시에
그 장소가 가진 새로운 가능성을 탐구하려는 목적이었다. 여기에는
빨간 풍선들의 가벼움부터 실제 건물에 쓰인 붉은 벽돌의 무거움까지,
변화에 대한 은유가 담겨 있다. 그럼에도 김이홍은 그곳에서 정성을 들인
'의식'과 같은 것을 열 필요는 없었다. 그의 의중을 짐작해보자면, 그는
계속해서 드로잉과 건축물 사이에 존재하는 극복 불가능해 보이는 간극을
실제의 구축과 허구의 선들 사이의 무언가를 통해서 계속 시험하고 있는
것처럼 보인다.

허구와 실재 사이의 대화

김이홍이 그의 건축적 공간 연구를 시작했을 무렵, 베아트리즈
콜로미나(Beatriz Colomina)는 2006년에 쓴 그의 에세이 〈미디어로서의
근대건축〉에서 이와 같이 주장했다. "건물은 이미지가 되고, 이미지는
일종의 건물이 된다. 그리고 이는 실제하는 건축 공간들처럼 사용된다.
(…) 건축의 모든 기념비적인 힘은 실체 없는 수단을 통해서 만들어진다."
이는 본 상에 직접적으로 해당되는 중요한 선언이다. 과거 다른 젊은
건축가를 위한 상들은 어느 정도 정해진 방식을 따라 두각을 드러내기
시작하는 신인을 발굴해왔다. 그들의 커리어는 보통 가상의 프로젝트들을
탐구하는 것으로 시작해, 여러 중요한 사무소에서 실무 경험을 쌓고,
마지막으로 자신의 아틀리에에서 설계한 작업을 짓는 것으로 완결된다.

그러나 허구의 공간 연구들(현상공모 목적이 아닌)과 실제 지어진
건축물들에 동일한 중요성을 둔 포트폴리오를 수상작으로 선정한
것은, 다른 이들에게 믿을 수 없을 정도로 큰 용기를 준다. '순간적인'
아이디어와 '실제의' 구축, 김이홍의 포트폴리오가 조명하는 것은 이 두
영역 사이의 일관성 있는 대화이다. 허구에서 실재로 향하는 일방향적
명료성 대신에, 여기에는 실재에서 허구로 향하는 반대의 움직임이
존재한다. 때때로 우리의 초기 작업들이 단지 최종으로 지어지는 작업의
습작에 불과하다고 가정하지만, 김이홍은 현실의 구축 작업들 역시
잠재적으로 개념 작업을 위한 연습이라고 믿고 있다. 두께감을 약화하고
얇음에 대하여 질문을 던지는 장난스러운 재료의 사용, 추상적 표현과
텍토닉적 표현에 대한 비교, 그리고 다양한 생각을 중첩해 만들어낸
복잡한 효율성에서 그의 이러한 생각이 증명된다.

비아르케 잉엘스(Bjarke Ingels)로 대표되는 '닷컴 세대'의 건축가들은
명확한 단계별 다이어그램을 건축을 위한 팝아트적 만병통치약이라고
강하게 믿는다. 하지만 김이홍은 공식적으로 포스트 닷컴 시대의

존 홍

건축가이다. 그의 세대는 건축물을 풍부한 복잡성과 다중 해석이 가능한 변화하는 언어로 인식한다. 전자의 세대가 1-2-3 단계를 거쳐 완결되는 디자인 과정이 주는 즉각적 희열감에 집착한다면, 김이홍의 건축은 지속적으로 스스로에게 의문을 던진다. 개념이 익숙한 유형과 매력적인 불안정성 사이를 끊임없이 움직이는 어떠한 형태로 재현되기 이전에 다시 반복하여 개념으로 돌아간다.

이러한 방식으로, 그의 작업은 오직 건축가만 이해 가능한 자율적인 게임에서 벗어난다. 허구와 실재가 대화하는 자리를 마련함으로써, 이 제스처는 일종의 공적 성격을 가지게 된다. 건물 사용자가 먼저 익숙한 감각을 느낄 수 있게 노력한 뒤, 그 익숙한 유형을 급진적으로 흔드는 공간 놀이에 방문자들을 참여하게 하는 것이다. 이는 허구와 실재가 서로를 부정하는 씁쓸한 자리가 아니다. 그의 57E130 NY 콘도미니엄에서의 진동하는 그리드나, DAN에서의 연속적인 보이드나 또는 코너스톤 1-532에서의 스케일을 무시하는 벽돌 곡면이나, 우리가 모두 이미 알고 있다고 생각하는 것들에 대한 판단을 잠시 유보하기를 미묘하고 유쾌한 방식으로 촉구하고 있다. 김이홍의 건축은 일상적인 것과 비일상적인 것 사이에 균형을 맞추면서 그 사이 틈을 만들어 우리 스스로 결론을 이끌어낼 여지를 마련해준다. 그의 건축은 건축이 무엇인가에 대하여 한 방향으로 가르치려들기보다, 그 복합적인 성격을 즐길 수 있는 자리로 우리를 초대한다.

영한 번역: 이영주 · 프로젝트: 아키텍처

존 홍은 건축가이자 서울대학교 교수로 재직 중이며 디자인 랩 '프로젝트: 아키텍처'를 이끌고 있다. 서울대학교로 부임하기 전에는 건축사무소 SsD의 공동 창립자(2004–2015)였으며, 하버드대학교 GSD에서 부교수(2006–2014)를 역임했다. 대표작은 『아키텍처럴 레코드』「메트로폴리스 매거진」「뉴요커」 및 「공간(SPACE)」을 비롯한 주요 매체에 소개됐고, 2014년과 2016년 베니스비엔날레 등 여러 국제 행사에 전시됐다. 또한 AIA 건축상, 「아키텍처럴 레코드」 디자인 뱅가드, 뉴욕 아키텍처럴 리그의 이머징 보이스 어워드 등을 수상한 바 있다. 저서로는 『융합하는 흐름: 한국의 현대건축 및 도시생활』(2012), 『새로운 주거방식의 조각들: 한국의 현대사회의 도시주거』(2016)가 있다. 그는 하버드대학교 GSD에서 건축학 석사를 취득했고, 버지니아대학교에서 건축학 학사를 취득했다.도시생활』(2012), 『새로운 주거방식의 조각들: 한국의 현대사회의 도시주거』(2016)가 있다. 그는 하버드대학교 GSD에서 건축학 석사를 취득했고, 버지니아대학교에서 건축학 학사를 취득했다.

Between Concept and Architecture

by Leehong Kim

1

I dreamed of becoming a carpenter during my middle school years. I still keep the two self-designed and handmade wood furniture in my room. I did not succeed in fulfilling that dream, but the carpentry workroom was an important starting point and a stepping stone for me to pursue my path as an architect.

Looking back, a carpentry workroom or a model room was always there in my work history, and I always did my studies and work in an environment that emphasized craftsmanship. For a year in graduate school, I did a part time job at the carpentry workroom in Harvard University's Carpenter Center for the Visual Arts which was the sole construction work by Le Corbusier in North America and built an architecture model for six months straight at Frank Gehry's office in Los Angeles. This was normal doing so as an intern because the environment there even had the partners to use models when making designs. Also, Steven Holl Architects in New York was also an office where design would be meaningless without models. For one project, three years were put aside to build models of various scales per design stage, and these models were used as references to the design. Because the models were not merely for visualization but to represent the actual building, various tools within the company were used to represent every single line and plane faithfully. This is what I enjoy the most: that is, not having the hands work independently from the head, but engaging myself in a process where the hand and the head resolve things in tandem. Different from the kind of work done in front of a computer, doing woodwork at a workroom serves also as a time of recovery, where I can focus calmly on a task that involves both the hands and the head. What I ultimately aim to accomplish through this process is to develop my thoughts to a higher level of perfection.

2

The work *Floor*, composed of a glass panel supported by the hands of approximately 180,000 plastic dolls of 5cm in height.

The wallpaper work *Who Am We?*, composed of approximately 200,000 circles that are smaller than one another by 0.4cm in diameter, resembling a pattern created by water droplets.

A 2.15m-tall armor titled *Some/One*, created by collecting approximately 70,000 elliptical US army dog-tags of 2.5×5cm in size.

A paratrooper titled *Paratrooper-1* that pulls at tens of thousands of strings with signatures of acquaintances embroidered on them to a singular point.

A handcrafted artwork that is composed out of countless repetition of objects tends to be impressive. There is a certain sublime beauty in Do-ho Suh's works that raise discussions between individualism and community as small objects are put together in his works to create an unexpected result at a higher level. Works that are created out of precise handicraft also tend to be visually appealing. Architecture is also born out of a constructive process that integrates labor and repetition.

In the 57E130 NY Condominium project, in which it was sought to express a sense of uniqueness with just the detail of a single brick tile within the homogeneous city system of New York City, much time was spent on reviewing the brick as the material. While existing as a brick building as a whole, the brick facade that is formed out of about 12,000 bricks and is embodying five different angles displays a distinct design while respecting the surrounding context of New York City.

From an interview, Do-ho Suh once said, "I wish to do a work that is easy to approach, but one that also has numerous transparent layers, such that one gets drawn deeper and deeper into it." As seen from his work that utilizes a semi-transparent Chinese thin silk, I wonder if his work process – that is, his precise handicraft – is that which not only projects the place where the work resides at, but also what that is beyond the work itself. The performance of his precise handicraft appears as an artwork by itself.

3

When I contemplate over physicality and non-physicality in architecture, I look up on the Beinecke

Rare Book & Manuscript Library of Yale University in the United States. It makes one wonder if it is built for human use, considering that it has an opening only on the ground level, with the rest of the 5-6 floors roughly finished with marble.

At times, this cuboid mass that is in alignment with the plaza's stone floor grid resembles a sculpture as well. It carries a completely opaque and meticulously measured appearance. However, a completely unexpected turn of events occurs when one looks inside. While having the same facade as the exterior, a sense of transparency and light-heartedness piques the visitor's curiosity. A feeling that cannot be expressed with words emanates from the interior space.

Space is an invisible non-physical element of architecture, and it may be said that it is the essence of architecture. Two essentially different substances – the exterior and the interior, and the physical and the non-physical – are divided by the 32mm thick marble panels. What appears as a mass of marble from the outside feels like a thin veil from the inside.

If transparency and opacity were put on a scale between 0 and 1, the two extremes can be manifested by the architect.

However, the countless degrees of transparency in between are achieved by adding a certain natural element onto a manmade background. The interior of this library takes on a different atmosphere and look according to the strength of light. A strong light reveals the intrinsic pattern and density of the marble, which creates an unexpected atmosphere. Standing at the boundary where the appearance of the physical material transitions into a non-physical space, one is struck in awe. The gap between physicality and non-physicality of architecture is definitely not as 'thin' as 32mm.

4

0.83% inclination is a very gentle slope that is hard to detect by walking. However, the moment when I discovered this inclination of 0.83% in a megapolis such as New York City, I was deeply awe-struck.

Of course, I learnt of the precise measurement later through a book: the incline rises by 18 inches (0.4572m) in height over a distance of 180 feet (54.864m).

The Lincoln Center is composed out of fragments of various exterior spaces. The plaza – which is surrounded by 3 performing theaters – is the most popularly visited spot amongst these exterior spaces. Every evening, the fountain at the center of the plaza, as though it is the main highlight of the Lincoln Center, creates music and performs ballet dances with its streams of water while being covered by fanciful lightings.

However, the thing that I'm drawn more towards is a shallow pool named as the Paul Milstein Pool of the Hearst Plaza located on the north side of the Lincoln Center area. Except for its size (approximately 42.6×19.2m), this placid and static-looking water surface that is at level with the ground level in the middle of a plaza covered with a plain stone pattern is not particularly eye-catching. One might even see it as the canvas of Henry Moore's sculpture *Reclining Figure* located in the middle. However, this pond's water surface that reveals that the plaza floor is at an incline of 0.83% asserts its state of existence in a reserved and elegant manner. With one side of the pond placed at about 15cm lower than the plaza, and the opposite side placed at about 15cm higher, the pond arrives to an equal level with the plaza at its middle point. It is where one can admire the detail of expression from the material aspect in an urban-scale project.

5

In June 2013 at MoMA in New York, a large-scale retrospective exhibition titled 'Le Corbusier: An Atlas of Modern Landscapes' was held. About 320 artworks including sketches, drawings, models, and photographs had occupied the entire floor of the museum, but the one that was most memorable for me was not his drawings or the models of a certain project, but a small scale bar drawn on the bottom right corner of the plot map for the Palace of the Assembly in Chandigarh, India.

The scale bar follows an ordinary format, but the dynamic look of the skating individual above the scale bar and the text '7 ½ MINUTES DE MARCHE' is a witty expression. It is certain that Le Corbusier's solution towards architecture and the space of megapolises must have begun from his investigation and interest towards human being. His attention towards human being is revealed even from the scale bar located on a corner of a blueprint. This scale bar informs us not only of the unit of distance but also the amount of distance based on human movement. A scale bar provides information that is usually more necessary for someone who is reading the blueprint than for someone who is designing it. However, this information was important to him even though he was a designer himself, and this makes one wonder how many more narratives at this meticulous human-scale level can be found in the spaces that he designed. Despite being a giant in the field of modern architecture who is still being regularly quoted in classrooms and workspaces, he demonstrates how it is the inhabiting human and not the space that should be the main focus of design.

6

M.C. Escher's fascinating drawings make one wonder if it is oneself who is moving or the picture that is moving. Although the drawing is on a 2-dimensional plane, the circulation that are expressed on it are not just on a flat surface. It is unclear from whence one should begin to study it, and perhaps it does not matter from whence one should begin to talk about it, as it all revolves to put the viewer back to the same spot. It feels as if one is being played by Escher, but it is not an unpleasant feeling. One becomes driven to find the reason behind a piece of his drawing, and ends up realizing that the effort is meaningless. Like a juggling performance in a circus, in Escher's works, the floor, wall, and ceiling exchange their roles and relationship rapidly, rejecting the boundaries of the XYZ axes and the sense of gravity and structure. As I design and contemplate always within the frame of gravity and structure, along with a division of the wall, floor, and ceiling,

Escher's works always come to me as a challenge. There are many projects that began with a design interest on the circulation. Even if the circulation that Escher had imagined are impossible from my perspective on reality, can it be possible from the user's perspective? Circulation are not only the representation of the users' footprints, but also the means for them to view and experience space. Through the circulation and the spaces, it is hoped that the users will come to experience things that were not even originally intended by the designer.

7

During my time as an intern in Los Angeles, I had an opportunity to visit the Death Valley National Park along with a visitor. It was a reluctant trip as I was not only unfamiliar with the area, but I was also not very keen to visit a desert. However, in contrast to my early expectations, I still have very vivid memories of that place.
Having one of the highest summer temperatures (highest record at 58.3°C) in the world and the lowest point in the western hemisphere (altitude at -86m), this national park possesses extreme environments. The Mesquite Flat Sand Dunes was especially a memorable experience. The desert scene created by wind in the early sunrise on the second day was a landscape that could not be created by human hands. The beauty was way beyond the lines that I drew out of deep contemplation. The curvy ridges that appeared like dancing waves that were formed undoubtedly by natural law were a continuation of incalculable lines. I liked and excelled in mathematics from my youth. I was even admitted into the university through mathematics. I also like graph papers. I am not obsessed with straight lines, but whenever I draw a diagonal or a curved line, the lines must always be drawn on a rational frame built upon mathematical calculations and coordinates. As someone who was bound to such laws, I felt so tiny when I was facing the sublime nature.

8
The Porsche is now synonymous with a sports car. Among its vehicles, the Porsche 911 has been a core reference model for over 50 years. 911 has a history of evolution over 7 generations. Not only

in terms of its exterior design, 911 has also adopted relatively unnoticeable changes in terms of the era-specific consumer demand as well as adoption of new technologies and materials. What is most impressive is the course of changes in the design over 7 generations. As presented in this poster, it is striking how different sections from 7 generations stitch seamlessly and harmoniously to complete a 911 model. Despite the changes in the models per generation by every 5-10 years, it displays an organic evolutionary process in the design's DNA over a long period of time.

Surely, the design was not done by one person over these 50 or so years. In this case, I think that it should be evaluated as a design of permanent value, and that a sense of permanence of design arises from a concept that acts as a solid foundation over the transitionary stages. Although there is a progressive evolution towards a more polished automobile over the course of time, ironically however, it seems as if the concept of the Porsche 911 model, instead of Porsche 911 as an automobile itself, is becoming more distinguished. This is also the reason why it is referred to with its specific name 'Porsche 911'.

9

I worked for 41 months at Steven Holl Architects from Jan. 2013 to May 2016, and participated in the design and construction administration of the Kennedy Center expansion project in Washington D.C. My daily encounter with Steven Holl and his attitude towards architecture influenced me greatly in forming my identity as an architect. Overlooking the Manhattan skyline, Holl's studio is where he makes his sketches, writes, and composes and imagines his projects. Similar sized sketchbooks are laid out in order on a deck above his table. The 4×6 inch sketchbook is something

Holl brings along wherever he goes – work, home, train, office, everywhere – to record down his observations and thoughts. His trademark watercolor sketches collect every information from 3-dimensional pictures to scaled detailed drwaings. I still remember vividly how he began each mornings with watercolor sketches, and how he handed them out to the project managers as he started work.

Being born in 1947, and having opened his office in 1977, his sketchbook must be at least 40 years old. His continuous and unchanging passion over that period of time must all be collected in those dozens of sketchbooks, and I believe that those sketchbooks provided the energy for Steven Holl to become who he is today.

PROJECT

DAN

As one climbs up the quiet and sloped Sinmunno-2-ga while admiring Inwangsan in the background, DAN project's slick appearance gradually comes into view. As a company building of a fashion brand, the identity of brand that had been conveyed from the very first meeting is unabashedly revealed. Also, in accordance with the will of the company, which was by no means light, the design was done with a focus on communication between the company and the customer as well as between co-workers.

The building mass that started out originally as a cuboid begins to be shaped in alignment with the setback regulation, main entrance from the road, outside view, and relationship with its interior. The communicative function was secured through the method of cutting out not just the exterior volume but also the interior mass. The deeply dug-out part on the third floor of the road-facing facade is where these cuts in the interior and the exterior come to coincide, and it is an important void that acts as a connection and allows communication between exterior and exterior, exterior and interior, interior and exterior, and interior and interior to occur. On the other side of the wall with the exterior finishing, external space is introduced along with the interior space. Also, there is a special moment where viewers from inside face the exterior wing-wall with exterior tile finishing and partial exterior view framed by corner openings of the wing-wall. By going up to the first floor from the front road, traversing across the corridor lined with round columns, and climbing up the external staircase to the third floor, one arrives at the new courtyard. While the external space of the front facade located at the corridor is open to all guests, this new external space on the third floor located at the building's center takes on the concept of a refuge for the staffs. It is not only a place where one can take a momentary respite while climbing up the external path that is connected from the ground floor entrance to the fifth floor, but also a space where staff members working in various floors can come to gather. The circulation that resembles the pathway of a sewing needle that penetrates in and out of the building provides interesting sceneries of the interior and the exterior from various viewpoints. The elevator and the external vertical staircase from the third to the fifth floors are composed as a single core to maximize the spatial efficiency of the interior work facility. The floors three to five where similar rooms are stacked above one another are mutually connected by an interior staircase which facilitates cooperation, communication, and movement among staffs.

Up to the 2nd floor level, the round column, the slab, and the frame structure that is finished with exposed concrete support the building mass that is finished with off-white tiles. Adding to the difference in material, the appearance of how the methodically positioned round pillars support the asymmetrical upper mass increased the discrepancy in terms of uniformity and forms. It was intended to retain the atmosphere of Sinmunno 2-ga by having DAN — which carries an imposing sense of existence in relation to its surrounding buildings — be finished with tiles that emit a sense of nostalgia.

Architect: Leehong Kim
Location: 1-137 Sinmunno 2-ga, Jongno-gu, Seoul, Korea
Program: office
Site area: 311.1m²
Building area: 156.69m²
Gross floor area: 835.52m²
Building scope: 5F, B1
Height: 16.86m
Parking capacity: 5

Building coverage: 50.37%
Floor area ratio: 195.97%
Structure: reinforced concrete
Exterior finishing: tile, Stuc-O-Flex, exposed concrete
Interior finishing: gypsum board, paint
Structural engineer: MOA & Consulting Engineers
Mechanical engineer: GIE
Electrical engineer: KS Engineering
Construction: YIGAK
Design period: 2016. 8 – 2017. 3
Construction period: 2017. 6 – 2018. 5
Client: 314HORNET Co.,Ltd
Photograph: Kyungsub Shin

57E130 NY Condominium

57E130 NY Condominium is a residential project that proposes a distinct design by introducing a change within the typology of row house in New York. It is comprised of a total of 5 units where each unit is equipped with six aboveground floors and one underground floor.

The city of Manhattan, which is elongated towards the north and south, is structured upon a grid system where it is divided into about 1,300 blocks. This project site is located at the East Harlem sector on the northeastern side of Manhattan, and it encompasses 18 feet and 3 inches (5.58m) in width and 100 feet (30.48m) in depth. It is the basic requisite of the district to adopt an attached wall composition and a continuous facade with the proximate buildings on both sides. As a result, most of the new construction projects in New York are mainly involved in planning a 2-dimensional facade than a 3-dimensional volume. Fortunately, because the proximate building facade in the project building's left and right were not equally aligned from the start due to the right building being retreated backwards for about two meters in distance, the right behind corner was already toting a 3-dimensional appearance. Along with the institutional restrictions, the facade context of New York City shaped the guideline for 57E130 NY Condominium. The general facade pattern where the opening becomes tripled by the four narrow bearing walls per floor was to be retained. With the brick module as its facing material, this project resolved the planning of the entire building mass and its specific facade details. The facade of this building was constructed with 27 bricks of 8 inches (20.32cm) each. Also, 40 bricks were used for each floor.

Instead of attaching a regular cuboid mass between the proximate buildings, the brick detail was used to express a mass that was fit into its adjacent buildings like a jigsaw puzzle piece. The lower left part (the left side of first two floors) and the upper right part (the right side of the sixth floor) was trimmed by about 1.5 times the brick's width at a 45 degrees angle.

In terms of details, a 12-inch (30.48cm) depth was added to the facade for an attempt to distinguish its appearance from the surrounding flat facade.

The pilaster (bearing wall) is 24 inches (60.96cm) in width and the window opening is 40 inches (101.6cm) in width. A half of the pilaster width is flat, while the other half is angled at 45 degrees to create a sense of depth. The direction created by the 45 degrees angle disrupts the overall consistency of the building and instills a sense of dynamism to the simple 3x6 facade grid. For the part where it was angled at 45 degrees, bricks shaped at 135 degrees angle were used. Abiding by the interior-specific conditions, two other parts with different widths in the pilaster were added and realized by bricks angled at 124 degrees, 146 degrees, and 158 degrees. 5 brick types with different angles in total were used to create the whole facade. The depth of the facade is emphasized even more by the shadows that it embodies.

Architect: Leehong Kim
Location: 57 East 130th Street, New York NY 10037 USA
Program: residential
Site area: 169.36m²
Building area: 107.17m²
Gross floor area: 746.71m²
Building scope: 6F, B1
Height: 18.29m
Building coverage: 63.3%
Floor area ratio: 343.54%
Structure: masonry structure
Exterior finishing: brick, stucco
Interior finishing: gypsum board, paint
Structural engineer: A Degree of Freedom
Mechanical engineer: New York Engineers
Electrical engineer: New York Engineers
Construction: Edgehill Construction, Inc.
Design period: 2014. 3 – 2015. 5
Construction period: 2015. 7 – 2018. 8
Client: Verse Development
Photograph: Hokyung Lee

Cornerstone 1-532

As a building that embodies the stimulating and disorderly diversity that arises from a combination of

the neighboring districts of Chugye University for the Arts and Ewha Womans University, the rapidly developing multiplex residential buildings, and the boundary between the luxurious residential region in the north and the commercial region in the south in a calm and serene manner, Cornerstone 1-532 was designed to be appreciated as the charm of Bugahyeon-dong in Seodaemungu.

With a building coverage of 40% on a 175m² sized land, the condition for a new construction was not favorable. However, the advantage of being located on the corner of a crossroad junction was utilized. By going for a continuous curved form that faces the road, and thereby minimizing the division within the building, it was attempted to maximize its perceived size. Another strategy was to utilize the elevation difference of the sloped land by raising the first floor level to allow natural light as well as a road entrance into the semi-basement floor.

The first floor was composed with a frame that was modelled after the site boundaries, and a new geometry that contrasts itself with the curved surface was inserted. Going beyond its role as a part of the building mass, the frame was also to be naturally connected to the wall by being extended as a canopy along the western side of the site where it touches the road as a part of an attempt to effect a building coverage of 100% in appearance. When approached from afar, a round stone-like mass states its monolithic appearance on the asphalt path; but from the pedestrian level, the frame functions as a medium of communication for the building's interior and exterior through the transparent glass fitted horizontally to it. A harmony between diversity and a contrast between texture and transparency is respectively achieved through the expression of serenity and stoutness of the building mass via colorless brick and the expression of dexterity and nimbleness of the frame via exposed concrete. The balance between opacity and transparency is also a kind of a compromise between the rental business interest of the building owner and the social responsibility of the architect.

The singular surfaces in the shape of a building are vertically stacked up from the basement floor to the third floor, and by adding a singular room at the back of the basement floor, a total of five rentable spaces are secured. The design was done without any knowledge of their future tenants, but the front room of the basement floor that resembles a semi-basement room and the space on the first floor have the potential to be wholly connected with the outside. The vertical mullian bar on the first floor frame was removed and was finished with a curved reinforced glass. This frame is also like a photo frame that captures the stories of what happens inside Cornerstone 1-532. Also, in the frame that takes after the form of the site, it is possible to enjoy the panoramic view of the Bugahyeon-dong beyond the curved glass surface.

Architect: Leehong Kim
Location: 1-532 Bugahyeon-dong, Seodaemun-gu, Seoul, Korea
Program: commercial (cafe, restaurant, artist studio)
Site area: 175m²
Building area: 69.75m²
Gross floor area: 310.62m²
Building scope: 3F, B1
Height: 9.7m
Parking capacity: 2
Building coverage: 39.86%
Floor area ratio: 118.28%
Structure: reinforced concrete
Exterior finishing: brick, exposed concrete
Interior finishing: paint
Structural engineer: Zywoo Structural Eng. Inc
Design period: 2010. 10 – 2011. 2
Construction period: 2011. 3 – 2012. 2
Photograph: Kyungsub Shin

Between Real and Fake: APMAP 2016 Yongsan

While the construction site for Amorepacific head quaters in Shinyongsan was busily ongoing with its frame construction stage in the summer of 2016, a three-floor temporary mock-up building was built on the northwestern corner before the main construction began. The installation titled *Between Real and Fake* that was installed in this 'fake' building provided a refreshing perspective to challenge the commonsensical beliefs that people have about buildings in general. There are so many things we take as granted as we stay within a building. This work, however, provides a new kind of experience by overturning our expectations. The structure where the installation work is located has the appearance of a building, but does not function as one. It has all the elements of a building – doors, windows, door handles, switches, lightings, CCTVs, and others – but they are not functional and are there only in appearance. By introducing a contrasting interior to this 'fake' building that looks real in the exterior, the building is revealed as a fake. As soon as one enters the entrance, one encounters two elevator doors. The installation work begins by pressing the elevator button, which leads the guests to the first entrance. As its introduction, instead of a comfortable elevator, the work forces the guests to climb up a steep staircase on a hot day. As the entrance is only reopenable from the outside, as soon as one has stepped into the entrance, one cannot but continue forward. Following the path, one is then led to the entrance to the second elevator. A momentary but an interesting mix of emotions takes place within the individual as one passes from a sense of curiosity when entering the door, to a sense of despair, and to a sense of relief as one is led outside. At the staircase where the sense of despair hits its climax, one is led to view the real building through an opening, through which one comes to intimately experience the difference between the authenticity and the inauthenticity of a building. Metal and orange colored PVC mesh were used as the main materials, and the orange PVC mesh forms a link with the orange elevator (reserved for construction purposes) of the proximate 'real' building.

Architect: Leehong Kim
Location: 100, Hangang-daero, Yongsan-gu, Seoul, Korea
Program: public art
Building scope: 405×420×500cm
Height: 5m
Structure: steel structure
Exterior finishing: steel plate, steel pipe, PVC mesh
Construction: Telim, Jongro Tent
Design period: 2016. 6 – 7
Construction period: 2016. 8
Client: Amorepacific Museum of Art
Photography: Jihoon Bae

Conceptual Work

Hide and Seek is a conceptual work that spatializes the psychological relationship between two

hypothetical protagonists. Both protagonists experience an inner conflict where each desires to become the other. By having each psychological state symbolized by an individual box, and by having the circular psychological state expressed as the state where a box in placed in a box, a box that repeats infinitely inside is thereby materialized. As it is minimally required to gather sides of three different axes to create space, the structural principle is to fit three units from the XYZ axes together to create space. These three units come from a 12×12 inch basswood panel. Each unit is built from a repeating process where a quarter of the panel is folded at a 90 degrees angle, and a quarter of the resulting panel is folded again in the same way, and so on. By fitting the three identical units together, a form of a box that continuously repeats itself is created. This infinitely shrinking box also hints at our inborn futility and vanity.

Labyrinth is a conceptual work that interprets and spatializes the relationship between Minotaur and Theseus, who feature in Picasso's 1935 work *Minotauromachy*. In ancient Greek myth, the hero Theseus, in order to defeat the monster-human hybrid Minotaur, enters the labyrinth and hunts the monster down. However, my interpretation of *Minotauromachy*, where one can see the two protagonists stretching their arms towards one another, is that the hunt goes both ways. This labyrinth is a space where the meeting cannot take place, and thus the paths of the two protagonists are materialized into a chain of two rings. The two protagonists are not in a same space, but are caught within their own spaces. They run around while sensing each other's existence through the transparent glass wall, but in this chain of two

rings, there is no spatial point of physical interception. With ten pieces of 25×2.5 inch-sized walnut wooden sheets and one piece of a similarly-sized acrylic sheet, the spatial design begins by moving and cutting out parts from the stacked mass of walnut sheets to create space. Three parallel passages are built along a long axis, and the acrylic sheet, which is inserted as a wall between the passages, acts as a medium of visual communication through its partial transparency. Minotaur and Theseus become locked up in a labyrinth where they cannot escape from nor hunt one another.

CRITIQUE

Between Fake and Real — Leehong Kim's Architecture and the Revealing of a Contemporary Dilemma

by John Hong (professor, Seoul National University)

I wish I could claim that the title 'Between Fake and Real' was my own invention. It is short and bold, but also contains within it the contemporary dilemma of young architects to stay true to the conceptual genesis of a project while negotiating the growing complexity of contingencies that conspire toward compromise. The title 'Between Fake and Real' is the resolutely uncertain position that architect Leehong Kim attributed to his APMAP 2016 Yongsan exhibit at the Amorepacific Headquarters site, a showcase of emerging artists that was staged while the Chipperfield building was still under construction.

Just as the title of his work portends, experiencing Leehong's exhibit firsthand was exhilarating in its banality. He first places us in front of a set of double elevator doors, a scene so familiar and yet so uneasy in terms of its ability to generate a sense of limbo and anticipation. We begin the sequence by pressing what looks like an elevator call button, repeating an everyday scene that we may have experienced countless time before. Once the LED indicator lights, we then enter the 'lift' on the left, only to find a surreal stair wrapped in orange construction fence mesh, a design move that references the still-under-construction Amorepacific building. But the generic material that usually wraps the exterior of a site to keep us out is inverted in its use; instead encasing an interior tube of space that we are meant to enter. Its misuse further mixes the everyday with the uncanny as the mesh filters the sunlight and creates a moire pattern out of the exterior view.

Following this stair up toward a bright aperture, we arrive at a semi-landing with slits framing fragmented views back to the Amorepacific construction site. Unlike normal stair landings, the space is not a place of rest but an extending conduit where we are forced to continuously move forward. Its curvature disorients us until we are finally led to a downward stair that ends up virtually where we started – the adjacent elevator door within the fake elevator lobby.

Both the title and the physical construction of 'Between Fake and Real' are powerful metaphors of Leehong's work. Through the use of irony, the actual physical artifact is also a paper-project, a theoretical statement prodding at the center of what defines architecture. It grafts into the way architecture, throughout its circuitous history, constantly flickers between reality and concept. Even as the contemporary critique of the fake in terms of endlessly reproducible references in architecture gained momentum through Jean Baudrillard's 1981 book, *Simulacra and Simulation*, the idea of creation and re-creation in and of itself has consistently haunted the 'authenticity' of architecture. The linguistic quality of architecture, its ability to act simultaneously as abstractly symbolic and physically embodied, can be traced back to the lineage of ancient religious structures that stood in as metaphysical dwelling places for the gods.

We sometimes forget that there have been many iterations of the fake throughout history, and it is perhaps time to bring the concept back into the foreground again to regain architecture's shrinking impact. Even the ancient Greek architects could not escape the fake: The use of the perspectival deceit of slightly bulging columns known as entasis haunts the 'purity' of their physical structures. However, it is the versions of the fake beginning in the 15th century with the Renaissance and accelerating into the 18th century 'neo' movements (as in neo-classical, neo-gothic, neo-baroque) that reveal the beginnings of a contemporary self-conscious deployment of artifice to achieve specific effect. Leapfrogging over modernism's implicit guilt for lost authenticity, in the later 20th century, representation through architecture launched into theories regarding pastiche, caricature, and collage. Whether we were left better or worse from postmodernism's pull, its questions remain as an essential, but unfinished discussion. That is why Leehong's choice of the word 'fake' rather than more indirect terms like 'simulacrum' or 'artifice,' is so important in his recognition as an emerging young architect. It provides insight into a potentially momentum-building aspect of the next generation of design: Fake does not hide behind academically obtuse debate. It is what it is and it is fake. But that does not empty fakeness itself from the intellectual conversation. Instead, it provides

a liberated state: fakeness is a conscious construct – it does not deceive the viewer but invites us in as part of the deceit. It is a kind of spatial playground we can collectively participate at different levels.

In this way, instead of aligning Leehong's work to the crises called forth by Baudrillard, the legacy of his architecture is more along the trajectory set forth by Palladio's Teatro Olympico, or more specifically by the way the theater was completed in 1585 by Palladio's disciple, Vincenzo Scamozzi. Forced into a truncated site within the footprint of an old fortress, Palladio distorted the perfect circular typology of a Roman theater to not only make it fit but to also create a new proximity between actors and audiences. But it was the later intervention by Scamozzi that completed the theory and reality of the project. His foreshortened stage, coupled with Palladio's foreshortened arena, compressed the perspective in a way that was obviously artificial to the audience but nonetheless communicated a powerful sense of depth. Scenographically, it connected the interior of the theater back to the exterior streets of the city. Likewise, as we will see in Leehong's work, the use of visual 'trickery' becomes a self-conscious interface between the urban scene and the individual, so that we can playfully engage with abstract concepts as they intersect with real-world constraints.

Between Material and Immaterial: Cornerstone 1-532

In Leehong's Cornerstone 1-532 project for instance, the design vacillates between fake and real, abstract and material, in several key ways. Occupying a small 175m² site, the legal building type for Cornerstone 1-532 is of the ubiquitous 'mixed-use' category that blankets much of Seoul, most of them created without the input of design architects because they are generic containers to fulfill a given building density ratio. To conceive of a project that is not just defined by the conditions of the site, Leehong cleverly leverages the corner lot line. Rather than accept its arbitrarily faceted boundary by extruding it to 'maximize' the area, Leehong transforms the building mass into a willfully gentle curve that brushes up against this line. The visual effect is of an ambiguously scaled surface where we cannot determine its end-point. Reiterating the property line's facets would have only emphasized the physical limits of the site, where this solution masks them.

However, the visual conceit does not stop here: by slightly extending the curve into a free-standing wing-wall on its far west side where the building must set back from the adjacent property, the curve masks the overall mass of the building. We are urged into believing that the building is wider than it actually is because this wall conceals the exact location of the back westerly facade as it meets the corner. Coupled with this hide and seek, there are no major windows in the curved surface, just the pixilation of bricks. From the steeply sloping surrounding streets, it is almost impossible to tell how tall the building is, rather at times, it looks like an inscrutable monolith.

Despite the building's heavy, drum-like quality, the ground floor is surprisingly cut horizontally and the void enclosed with delicately curved transparent glass, defying the apparent weight of the brick above it. The two materials resonate dialectically, contradicting and forcing the hand of the other: The immaterial glass, assembled with vertical silicone joints and supported by hidden top and bottom mullions, is the playful opposite of the textured and crafted bricks above them. The former denies its tectonic properties while the latter reinforces them. Together they represent the real and fake: It is as if the glass void of the ground floor reveals the visual game of the brick curve as both material and metaphor. Finally, instead of brick and glass directly meeting, a concrete lintel concisely expresses the between-ness of real and fake: This thin line, instead of curving like the brick or glass surfaces, obediently registers the faceted site boundary as if to place a real measure of the conditions around which the unreal dance.

Between Thick and Thin: DAN

It is of central importance that even before Leehong started designing his own buildings, he engaged in his own self-directed spatial studies at the model scale while working at other offices. These models, or we could even call them constructions, were large and meticulously crafted, the scale between the ephemerality of drawing and the tectonic certainty of a building. While his peers were engaging in open competitions trying to play the gamble of the 'big win,' Leehong was quietly developing and tuning his conceptual skills, as if exercising for an unknown future event.

For instance his Hide and Seek study from 2011, built while he was working at the corporate-scale Korean office SAMOO Architects & Engineers can be directly seen as a precursor to the fashion brand headquarters DAN completed in 2018. The model explores the clever conceit of the thin plane by folding four sheets of flat material to additively generate a cubic volume that appears as if it was instead made from thick subtracted volumes. The exercise has no program, only potent architectural implications. Instead of rushing toward a singular solution, it is a kind of delayed spatial research that can be deployed across a wide variety of future situations. Most importantly, it is a work-in-progress: an exercise that can be repeated and refined in parallel to an actual finished work.

Part of the implicit existence for the Korean Young Architect Award is to recognize the successful carrying out of ambition: Only seven short years later in 2018, Leehong implemented his first version of 'Hide and Seek' in the DAN building. Using the same method of folding surfaces, the project is organized around a series of subtracted voids in plan and section. These voids are registrations of the human body and lines of sight as they tunnel through the imaginary overall mass. Beginning from the facade, a stair rising a half-floor from the parking area seems to carve out a double height void on the left side of the building. This void then follows across the front of the building, defined by a colonnade that frames the main entry to the fashion offices. The void then burrows up the right side of the building to the upper levels. The 'inefficient' circulation that forces one to cross from one side of the building to the other is of course dictated by the complexity of topography, zoning constraints, and building setbacks. Nonetheless, the playful materialization of this practical reality engages the visitor, inviting us into the conceptual conceit.

Notably, a deep, enigmatic void on the third floor leads us upward in a kind of 'hide and seek' to find its rationale. When we finally reach it, we are doubly rewarded through the realization that it is not only a void cut horizontally into the facade but also downward back to the first floor colonnade. It is a kind of conceptual wormhole, a direct and visual shortcut between entry and upper levels, even as the body itself must traverse a long path to get there. Although in the plan drawings this void appears insignificant, it is a key conceptual moment where we realize that the thickness of the building's subtractive language is in fact made of thin folded surfaces conceived exactly like Leehong's Hide and Seek model. Like in Cornerstone 1-532, part of the facade next to this void becomes a free-standing wall that loosely frames exterior space, amplifying the exterior wall as both thick and thin. The materiality of the facade reinforces this fakeness: The ubiquitous architectural tile one might see throughout Seoul is cut in half so its new micro-dimension unsettles its generic quality: As it 3-dimensionally wraps the voids, we question if it really is the everyday tile we all think we know.

From this third floor, the stair rises up through the fourth and fifth floors, burrowing a new void into the narrow body of the building in response to the zoning setbacks. Even within a small site in a high-priced real estate zone, Leehong still manages to insert several key slices through the ceiling of the third to fifth floors culminating in a skylight that spills natural light into the work areas. Even as these voids are narrow, they create dramatic subtractions and cracks of views that produce the illusion that the small building footprint is bigger than it actually is.

Following this crafted journey that reveals a consistency in design logic, the play between void and surface is consciously undermined in an ultimate moment on the fifth floor. As the building rigorously operates within the logic of subtraction, a single additive moment at the scale of one person cantilevers from the stair landing. From this visceral vista, we can not only look over the city but also gaze back at the sequence of tunneling voids: It is as if the architect is revealing that this is all a game while inviting us to step outside of the visual conceit for a brief moment to examine and even analyze the project's spatial operations.

When exiting the building and returning to the ground level of the city, the first floor colonnade can also be discerned as a formal game that risks a close encounter with the post-modernism of the 1980s. A fake concrete column with no support below it, constructed exactly with the tectonics of the two real columns that flank it, is inserted to obey the familiar rhythm of the colonnade typology. This one seemingly insignificant moment reveals the sleight hand of the architect. It is a definitive statement that architecture is first and foremost a linguistic game, not a tectonic endeavor. However, unlike Eisenman's (in)famous fake column at the Wexner Center for the Arts that dramatically calls attention to itself as it hovers impossibly over a stair, Leehong's column dissolves into an expected typological background, only noticeable if you carefully look for it. In this way, Leehong's language is not about disruption, but about a gently prodding transformation. Before we can reject his game, we are led to become immersed in it.

Between Labyrinthine and Mundane: 57E130 NY Condominium

One of Leehong's most recent projects is a narrow sliver-like townhouse condominium in Harlem, New York. Just as in his other built works, the actual building is inextricably linked to one of his fictional self-directed studies, Labyrinth, conducted in 2010. In this earlier exercise, he explored the simple stacking of a module to generate a complex series of spaces that intertwine and pass through each other while never actually touching. Made of scaled-down wood bricks, Labyrinth, is another one of Leehong's enigmatic 3-dimensional models that are materially visceral, but are, in the end, conceptual exercises in spatial play. Furthering the efficacy of the study, Leehong created his own set of constraints: An imaginary boundary elongates the artifact, even as there was no actual site to contend with.

The extended form of Labyrinth has striking similarities to the New York townhouse that Leehong would design and build eight years later.

New York zoning and building codes are notoriously some of the most stringent in the world. They are further exasperated by zero-lot line sites where buildings literally share the structural walls between them. This not only makes construction extremely difficult but also aggravates solutions to meet the legal requirement for light and air. Further complicating the issue, this particular site was only 5.5m wide: like his Labyrinth study, the long and narrow footprint challenged the integration of five housing units, each with separate private entries. To top off the complexity, a large elevator that fits the New York stretcher law for transporting a person in the prone position and a code-compliant egress stair with the same standard requirements found in larger buildings created impenetrable obstacles that made a 'logical' plan impossible. In essence, the difficult program requirements were labyrinthine.

In a similar way that Leehong cut multiple layers through his Labyrinth object to reveal the complexity of the spaces in plan, the Harlem townhouse reflects a similar puzzling together of the program and circulation. Even as developer driven designs of this building type have an inevitably blunt quality in an effort to minimally satisfy the legal codes, in Leehong's version there seems to be an interlocking spatial logic that reflects the brick module of his precedent theoretical study. With the corners of rooms missing, we realize that the operational scale is not the room itself, but a small and rigorous pixelized overlay that allows for the solving of storage, light, entry, egress, and quality of space. In the end, where other architects might fear the lack of a clear, diagrammatic organization, Leehong seems to use the motif of the labyrinth to embrace the complexity that it brings. In other words, instead of allowing a concept of 'efficiency' to limit the solution, the difficulty of the plan is used as a spatial generator to actually raise the overall quality of the interiors. Here is where the mundanity of building codes and the surreality of labyrinthine space can converge.

Going back to brick, Leehong uses the material in a literal sense in the cleverly conceived facade. Even as the apertures in his building pick up the repetitive and self-similar punctuation of windows seen in the adjacent urban fabric, he thickens the openings through a chamfered assembly of brick modules. The result is that the grid of windows appears to subtly shift as the chamfers are applied to one side of an opening or the other. This folding of the surface makes the substantial brick seem paper-like at times, while simultaneously creating an overall effect of depth. Just as in previous projects, the design move is not dramatically discernable, and therefore is initially felt rather than analytically perceived. Rather than announcing a stark difference, the boundary between fake and real is a complex back and forth: We are first invited to gaze at the proportioned regularity (and even mundanity) of the facade design, before being compelled to question its unsettling alignments. Upon further experience, the facade's heavy materiality becomes a foil for ephemeral play: If the building was made out of something light, its instability would not be as significant.

As an interesting end-note, before the construction began, Leehong installed a temporary exhibit of ascending red balloons on the site. Called *Harlem Celebration*, it acted as a kind of funeral service for the dilapidated building that once occupied the site, calling it out as a place for future potential. From the lightness of the exhibition's red balloons to the heaviness of the actual building's red bricks, the metaphor of transformation is engaging. Nevertheless, Leehong did not have to host such an elaborate 'ceremony' on the site. If we are to psychoanalyze his intentions, he seems to constantly want to mark the impossible gap between drawing and architecture with something in between the real construction and the imaginary lines.

A Dialogue between Real and Fake

Around the same time Leehong began making his architectural spatial studies, Beatriz Colomina in her 2006 essay, 'Media as Modern Architecture' claimed, "Buildings become images, and images become a kind of building, occupied like any other architectural space... All the monumental force of architecture is generated by the most insubstantial means." This is a significant statement that can be directly applied to this award. In the past, other Young Architects Awards perhaps looked to emerging professionals that were following an accepted path: Their careers began exploring hypothetical projects, to working in significant offices, to finally culminating with built work in their own ateliers.

I find it incredibly heartening, however, that this award program selected a portfolio with equal footing in both fictional spatial explorations (that are not competitions) and actual buildings. Most significantly, instead of reinforcing an idea of linear progress from 'ephemeral' ideas to their 'actual' materialization, Leehong's portfolio expresses a constant dialogue between the two realms. Instead of a clear forward motion between fake to real, there is a reverse backtrack from real to fake. Where we sometimes assume that our earlier works were just rehearsals for the final built product, Leehong believes that the real is potentially also a study for the fake. This is evidenced by his material play where thinness questions and undermines thickness, where tectonic expression is measured against abstraction, and where efficiency is complicated by the overlay of multiple ideas.

Where 'dot.com generation' architects typified by Bjarke Ingles insist on the clear inevitability of step-by-step diagrams as a kind of pop-art panacea that can solve all of architecture's woes, Leehong is officially an architect from the 'post-dot.com' era. His generation sees architecture as a transformative language with rich complexity and multiple-readings. Where the former fetishizes the immediate gratification of the 1-2-3 start-to-finish design process, Leehong's architecture constantly prods itself, backtracking through the iterations of ideas before re-presenting them again within forms that flicker between familiar typology and fascinating instability.

In this way, his is not some version of an autonomous game that only other architects can understand. By placing the fake and the real in dialogue, it is a kind of public gesture: an effort to invite the building's users to first occupy a familiar scene before engaging them in a spatial game that progressively destabilizes the typology. This is not a bitter process of negation, however. Like his vacillating grid in 57E130 NY Condominium, the connected voids in DAN, or the scale-defying curved brick in Cornerstone 1-532, they are subtle and sometimes humorous urgings for us to suspend our disbelief about what we think we already know. By playing the balance between the mundane and the extraordinary, Leehong's architecture creates a state of in-between-ness that generously gives space for us to make our own conclusions. Rather than forcing us into a didactic understanding of what architecture is, we are invited to enjoy its multiplicity. (Translated by Keunho Hong)

John Hong AIA, LEED AP is an architect and professor at Seoul National University and the director of design lab Project : Architecture. Prior to joining the faculty at SNU, he was co-principal at the critically acclaimed firm SsD (2004–2015) and professor in Practice at the Harvard GSD (2006–2014). His work has been exhibited at international venues including the 2014/2016 Venice Biennale and published in major media such as *Architectural Record*, *Metropolis Magazine*, *The New Yorker* and *SPACE*. His built work has received fifteen AIA awards, *Architectural Record*'s Design Vanguard, the Emerging Voices Award from the Architectural League NY, to name a few. His most notable writings include the books, *Convergent Flux: Contemporary Architecture and Urbanism in Korea* (2012). He received his Master's in Architecture with Distinction from the Harvard Graduate School of Design and a Bachelor's in Science in Architecture with Honors from the University of Virginia.

문주호
임지환
조성현

경계없는작업실

Jooho Moon
Jihwan Lim
Sunghyeon Cho

Boundless Architects

문주호
대한민국 건축사이며
경계없는작업실의 파트너이자
대표이다. 서울대학교 건축학과를
졸업하고 경영위치건축사사무소에서
실무를 쌓았으며 경계없는작업실을
열어 열린 사고를 바탕으로 공간
가치의 진화를 고민하고 실천하고
있다. 현재 서울대학교 건축학과에
출강 중이다.

임지환
대한민국 건축사이고,
경계없는작업실 자문파트너이자
제로투엔의 대표이다. 서울대학교
건축학과를 졸업하고 한울건축에서
건축 실무 경력을 쌓았고
경계없는작업실을 열었다.
경계없는작업실에서 공간개발과
기획-설계-시공의 통합가치
필요성을 느끼고 종합건설업을 가진
제로투엔을 창업해 공간의 가치를
늘리는 다양한 작업을 이어가고 있다.

조성현
경계없는작업실의 파트너이자
스페이스워크의 대표다. 서울대학교
건축학과를 졸업하고, 아이아크에서
건축 실무 경력을 쌓은 뒤
경계없는작업실을 열었다. 현재
팀의 사업개발을 담당하고, 사내
기술팀인 Boundless-X가 독립해
창업한 스페이스워크를 통하여
다양한 전문가와 기술에 기반한
건축, 도시 문제 해결에 집중하고
있다. 서울주택도시공사에서
도시재생자문위원과 외부자문위원을
맡아왔다.

Boundless Architects
"Boundless" is a group of architects
based in Seoul, Republic of Korea.
It is an architecture-based broad
church(open organization) trying
to make innovations by searching
gaps among various boundaries.
"Boundless" offers the city new
spaces through constant study of
architecture.

Joonho Moon, partner
Jihwan Lim, partner
Sunghyeon Cho, partner

에세이

경계 찾기

문주호 임지환 조성현

문주호 임지환 조성현

I

건축가가 되고 싶다는 어릴 적 막연한 꿈은 건물에 대한 관심과 호기심에서 비롯됐을 것이다. 건축가로서 지금 우리는 다양한 맥락으로 건물을 판단하지만 20년 전에는 그저 가장 높은 63빌딩과 시카고의 시어스 타워를 동경했다. 대학입시를 준비할 때도 벼락치기로 공부한 가우디보다, 사실은 두바이에 새로 지어질 부르즈 두바이와 도쿄도청사와 63빌딩의 높이가 더 궁금했다. 건축과에 들어와서는 친구들과 아파트 층 수와 세대 수를 비교하고 입면에 보이는 발코니 베이 수를 세어보며 넓은 평형의 웅장함을 열망했다. 그때 우리에게 건축의 가치는 기술의 결과였고 부와 성장의 상징이었다.

그런데 얼마 전 완공한 국내 최고층 타워인 롯데월드타워는 흥미로우나 감명을 받진 않았다. 대중 역시 더 이상 건물의 규모와 높이에 예전만큼의 흥미를 느끼지 않는 듯하다. 오히려 서울시청사와 DDP의 디자인에 관해 더 많이 논의하고 좋은 공간은 핫플레이스로 앞다투어 소개되어 더 많은 사람이 공유하고 경험한다. 우리가 앞으로 탐구할 공간에 관한 이야기는 건축가인 우리의 이야기이자 건축을 경험하는 모두의 이야기이기도 하다. 결국 우리 역시 같은 시대 속에서 함께 살고 느끼고 고민하는 사람들이다.

문주호 임지환 조성현

2

우리는 다이어그램, 이미지, 도면, 설명글을 통해 건축물의 의도와 의미를 설명하고 설득한다. 건축주와 수많은 회의를 거쳐 건물의 방향성을 결정하고, 팀원 간에도 고집과 주장, 설득의 과정을 거쳐 하나의 건물을 완성한다. 그리고 우리는 서로에게 물어본다. "그래서 이 건물 예쁜가요?"

건축물의 아름다움, 그리고 공간감은 논리의 영역이 아닌 취향의 영역이라 한다. 그렇기에 며칠 밤을 고민해서 만든 자신 있는 결과물도 디자인이 마음에 들지 않는다는 건축주의 한마디로 세상에서 사라지고 만다.

그럼에도 우리는 아름다움을 취향의 영역으로 단정하지 않는다. 진부한 예시이지만 안도 다다오(安藤忠雄)의 빛의 교회는 건축을 시작했던 대학 시절 우리에게 본능적 감동을 주었다. 바르셀로나 여행에서 우연히 발견한 몬주익 언덕 뒤편의 전망은 다른 사람들에게 적극 추천하고 싶은 장소이다. 숲을 거닐다 쉬려고 누운 곳에서 바라본 나무와 하늘의 모습은 마냥 행복하다. 관광명소 중에는 더러 허울뿐이거나 인문학적인 맥락만을 가진 곳도 있지만 사랑받는 장소들은 여전히 많은 사람에게 공감을 주고 사람들의 감정을 공유하게 만든다.

그렇기에 우리는 공간이 당연히 그래야 하는 것에 대해 항상 질문을 던지려고 한다. 공간의 본질적 가치에 대해 질문하고 그 가치를 경험하는 사람의 가치관에 대해 끊임없이 대화해 우리의 건축적 자산으로 축적하려 한다. 단순하게는 집은 비가 새지 않아야 하고 볕이 잘 들어와야 하며 따뜻해야 한다. 우리는 우리의 관점을 주장하기 이전에 당연한 것을 당연히 만들 수 있는 실력을 가진 건축가이고 싶다.

3

그럼에도 우리가 생각하는 건축은 본능적 가치를 넘어 시대와 도시의 시공간적 맥락의 무수한 관계 속에서 찾아낸 균형점의 결과라고 생각한다. 우리가 고민하는 요소는 공간감, 건축주의 목표뿐 아니라 그 자리에서 오랜 시간 테두리를 만들어온 땅의 누적된 관계를 보아야 하고, 건물이 지어지고 난 뒤 연속될 새로운 관계의 요소들에 대해 이야기해야 한다.

표현이 거창했지만 결국 우리의 작업은 길과 건물 사이의 관계, 주변과의 형태적 조화, 창의 적절한 위치, 건물의 수익성과 사용성, 그리고 주변 건물에 끼치는 영향 등을 우리만의 방법으로 고민해 공간적 해법으로 제시하는 것이다. 그래서 우리는 기술과 상상력을 기반으로 최대한 많은 것을 고려하고, 그것이 결과로 이어질 수 있도록 도전한다.

얀 포르만(Jan Vormann)의 작품을 보면 오랜 시간 동안 부서지고 변형된 건물의 빈틈을 다양한 레고 블록들을 조합하여 원형의 외곽선에 가깝도록 재조합했다. 그 모습은 OMA의 요코하마 다이어그램 등 건축가의 공간 분석 다이어그램과 닮아 있다. 레고 블록 하나는 우리가 공간에 반영해야 할 다양한 주제이며 그 모습은 한 개인의 의도가 아니라 맥락의 경계선과 다원적인 요소들의 결합으로 만들어진다. 우리가 추구하는 건축의 방향성 또한 이와 다르지 않다. 경계없는작업실은 건물 외형으로 드러나는 일관된 색깔을 지양한다. 끊임없이 관점을 확장하고 고정관념을 넘어 그곳에 필요한 다양한 모습을 만들어간다.

문주호 임지환 조성현

4

취미와 특기를 명확하게 말하긴 어렵지만 무언가를 목적 없이 기꺼이 하는 것이 취미라면 우리에겐 지도 보기가 될 수 있다. 지도를 보며 국내와 세계 곳곳을 유심히 살피다 보면 당연하다고 여기는 것과 새로운 장면을 왕왕 발견하게 된다. 우리의 풍경 색은 전통적인 기와처럼 수묵화의 무채색과 감색이다. 그런데 위성지도를 멀리서 보면 꼭 기와뿐 아니라 건물이 모여 있는 도시의 색, 숲의 색이 전체적으로 무채색과 감색의 느낌이다. 스페인이나 포르투갈을 멀리서 보면 붉은 토양의 색인데 도시 스케일에서 관찰하면 붉은 기와를 가진 집들과 재료들이 도시의 풍경을 붉은색으로 만든다.

삼국지에서 계륵이라 불리는 도시 한중은 촉의 근거지인 쓰촨성과 장안의 중간 산맥에 좁게 형성된 분지지형이다. 두 거점으로 이동할 때는 중요한 전략적 통로이지만 두 도시와 비교할 때는 상대적으로 계륵이라 불릴 만큼의 크기를 가지고 있다. 천안은 호남과 경남으로 가기 좋은 지정학적 위치를 갖고 있으며, 화산대에 위치하는데, 지진이 만든 도시들은 지각판들의 충돌이 만들어내는 특유의 지형을 가진다.

이렇듯 위성지도를 통해 관찰하면 짧게는 수백 년, 길게는 수만 년 이상의 시간을 만날 수 있다. 그 오랜 시간 거대한 지형적 공간 사이에서 수많은 사람이 삶을 개척하고 도시를 만들어왔다. 상상할 수 없는 크기의 도시가 만들어내는 이야기 속에서 우리 역시 독립적인 존재가 아닌 연속된 도시의 시간 중 일부로 이야기를 덧붙여가고 있다. 우리가 건축을 한다는 것은 수많은 건물 속에서 고작 하나를 만드는 일이기도 하지만, 우리의 시간을 연결하는 일이기도 하다.

5

지도 관찰과 마찬가지로 도시 공간 탐험은 우연한 만남과 즐거움을 준다. 여행을 할 때면 도시 하나를 체류일 수로 나눠 하루에 한 지역씩 무작정 걸어 다닌다. 어찌 보면 무식한 방법이지만 도시를 여행하고 관찰할 때 이만한 방법이 또 없다. 이렇게 걷다 보면 여행 잡지나 지도에서 경험할 수 없는 도시의 날것들을 볼 수 있다. 이름 모를 건축가의 작업들을 발견하는 것도 기분 좋은 일이고, 수많은 사람이 만들어온 도시의 생생한 모습은 더 큰 감동을 안긴다. 다양한 지역주민의 삶과 철학, 그리고 일상 속에서 오랜 시간 쌓여온 도시의 건축은 그 자체로 커다란 공간적 경험을 선사한다. 마을이 가진 날씨와 지형, 그리고 사람과의 관계가 만들어낸 공간은 우리가 경험하는 일상의 고정관념에 수많은 질문을 던진다.

특히 쿠바 여행은 인상 깊은 경험이었다. 수치상으로는 가난한 나라이지만 아바나에서 우리는 "그래서 너희는 행복한가?"라는 질문 세례를 받고 온 듯한 느낌이었다. 호텔이 거의 없어 일반 가정집이 숙소 역할을 하는데, 허름해 보이는 외관과는 달리 내부는 호화롭지 않아도 강렬한 색감과 구성을 가지고 있었다. 우리는 최신의 핸드폰과 텔레비전을 가지고 있지만 그들은 여유로운 시간과 낙천적인 삶의 태도를 가지고 있다. 이런 그들의 모습은 공간 구석구석에 자연스럽게 녹아 있었다. 깊숙이 들어가면 그들 역시 우리의 일상을 부러워할 수 있지만, 이러한 도시 경험 방식은 내가 가진 공간을 다시 질문해볼 수 있는 좋은 기회다.

6

도심 한가운데 항구가 있는 부산의 풍경은 역동적이다. 특히 항구의 컨테이너와 크레인 그리고 배가 만들어내는 거대한 스케일은 문명의 성취에 대한 경외감마저 든다. 하지만 우리가 항구 풍경에 매력을 느끼는 가장 큰 이유는 역설적으로 중량감과 기계장비의 움직임이 만들어내는 가벼움에 있다.

건물 크기의 컨테이너와 배들이 목적지와 시간에 따라 움직이며 해안가의 풍경을 다채롭게 한다. 컨테이너의 다양한 컬러와 수량이 리듬감을 만들고, 배와 크레인의 움직임이 변화를 만들어낸다. 그렇다면 우리가 만드는 건물은 어디까지 가벼워질 수 있을까? 늘 30년을 내다보며 건물을 지어야 할까? 내부 쓰임새나 인테리어가 달라지는 것 말고 정말 건물 그 자체의 물리적인 환경이 변화할 수 있을까? 그리고 도시는 이러한 변화를 필요로 할까?

단정할 수 없지만 오늘날 많은 공간은 점점 더 가벼워져야 하고 변화에 대응할 수 있어야 한다. 공급이 수요를 넘어서고 콘텐츠의 집중과 이동은 점점 다양해지고 빨라지는 것에 비해 건축은 여전히 무겁다. 예전에 비해 리모델링과 인테리어 문의가 늘어나고 있기는 하다. 하지만 여전히 많은 건물이 사용되지 못하고 용적률 및 용도와 맞지 않아 사라진다.

얼마 전 성수동 근처 식당에서 식사를 했다. 인테리어가 세련되어 자세히 살펴보니 예전 카페 인테리어를 고치지 않고 그대로 쓰고 있었다. 공사장 인부가 대상이라 철거비조차 아낀 것일 수도 있지만 공간적인 가벼움을 경험한 느낌이었다. 우리 역시 세드릭 프라이스(Cedric Price)의 'Fun Palace'와 같은 가볍고 변화하는 건축을 일상에서도 만들고 싶다.

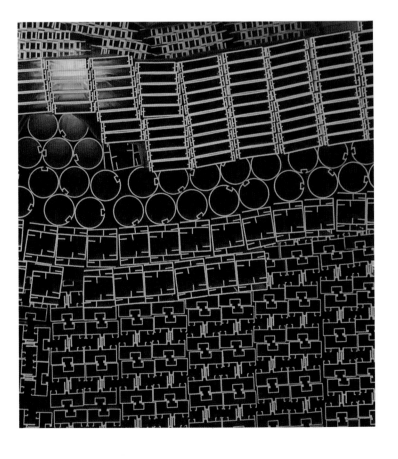

7

도시와 공간에서 이러한 고민들을 실천하기 위해 우리는 법규와 비용, 기간이라는 세 가지의 현실적인 질문에 답해야 한다. 건축은 많은 자본이 필요하고, 자본이 있더라도 건축을 경험하는 대다수의 사람은 집을 짓다가 10년은 늙는다고 한다. 땅을 알아보고 건축가를 섭외하며 각종 금융비용과 세금처리 그리고 운영까지, 이어지는 일들은 건축가인 우리에게도 어려운 일이다. 게다가 정보는 여전히 폐쇄적이고 사람들은 발품을 팔며 부정확한 판단 속에서 많은 비용을 투자한다.

그렇기에 아파트라는 주거 형식은—아파트 단지의 크기와 도시가로에 대한 폐쇄성 등 부정적인 부분은 많지만—한국의 현실에서 객관적으로 매력적인 상품이다. 기본적인 주거의 기능을 충족하며 표준화된 상품으로서 거래가 활발해 자산의 가치로서도 매력적이다. 르 코르뷔지에가 한국의 아파트 풍경을 어떻게 바라볼지는 모르겠지만 기능적으로 그가 생각한 주거의 미래가 한국에서 가장 근접하게 실현되고 있다고, 모두가 생각할 것이다.

우리는 아파트를 예찬하는 것이 아니라 아파트를 소비할 수밖에 없는 상황을 파악하고 앞으로의 건축에 적용해 변화를 만들어내고 싶다. 작은 현장일수록 사람들의 손으로 만들어지기 때문에 비용과 시간은 예측하기 어렵다. 하지만 냉정하게 이야기해서 세상에 이런 상품은 없다. 불확실성의 부채를 모든 관계자가 불합리하게 나눠가지는 구조이다.

기술의 발전을 이용해 양질의 공간을 더 많은 사람에게 제공할 수 있는 합리적 체계를 만들어야 한다. 좋은 공간을 위한 창의성과 합리적 상품은 결코 이율배반적이지 않으며 많은 사람이 더 많은 정보를 가질수록 궁극적으로 건축문화가 진화할 수 있을 것이다.

8

서두에서 밝힌 것처럼 현재 우리는 여러 부분에 도전하고 있다. 하루에도 몇 번씩 이러한 관점에 대해 고민하고 현실적으로 어느 하나에 집중해야 하지 않을까라는 질문을 던진다. 한편으로 앞서 이야기한 것들이 하나라도 빠지면 좋을 공간일 수 있을까 하고 또 되묻는다. 좋은 공간을 만들기 위해 생각을 더 넓게 확장하고 싶은, 아직까지 우리는 열정 넘치는 젊은 건축가이다.

어느날 카페에서 작업 중인 내 모습을 보고 나도 모르게 사진을 찍었다. 분명히 기술이 발달해 내 가방은 가벼워져야 하는데, 챙겨온 물건들은 작업할 때 없어서는 안 될 물건들로 가득 차 예전보다 더 무거워져 있었다. 우리가 경험하는 시대는 어쩌면 거대한 변화의 과정 속에 있으며 우리는 그 경계를 살아가는 경계인이지 않을까.

한참 설계자동화를 개발하고 있을 때 알파고와 이세돌의 경기가 사회적 이슈였다. 우리의 방향이 시대의 흐름 속에 있다는 것을 확인해 기쁘기도 했지만 한편으로는 거대한 변화가 만들어내는 새로운 세상의 모습이 두려웠다. 하지만 우리는 시대 변화 속에서 더 좋은 공간을 더 편리하게 만들겠다는 목표를 갖고 창의성과 기술을 결합해보고자 한다.

9

우리는 어떤 건축가를 꿈꿨고 어떤 건축가이며 어떤 건축가일 것인가.

이제 막 30대 중반을 넘은 '젊은' 건축가로서 우리의 건축관에 영향을 준 현상들에 관해 명확하게 정리해 이야기하기엔 아직은 너무 젊다. 건축을 통해 세상을 변화시키고자 하는 열망은 가득하지만 우리는 이제 겨우 밑그림을 그리는 과정 속에 있다. 지금 이 글을 쓰는 순간에도 우리의 건축 세계에 영향을 주는 수많은 현상과 경험이 변화하고 있다. 에디터에게 마감을 약속한 2018년 7월 10일의 경계없는작업실과, 부랴부랴 마감을 실제 준비하고 있는 2018년 7월 29일의 경계없는작업실 또한 다르다.

세상에 우리의 방향성을 주장하기 위해 과정을 생략하고 사고를 축약하여 선언하듯 이야기하는 지금의 단어들은 하나의 목표이기도 하지만—사실—아직은 전체를 보여주기 어려운 부피의 단면도 한 장에 불과하기도 하다.

하지만 확실한 것은 우리는 변화에 열려 있으며, 지극히 현실적이다. 또한 관찰과 발견을 좋아하고 새로운 관계들을 통해 예측하지 못했던 방향성과 복잡성을 지향한다. 궁극적으로는 이러한 시도들의 결과로 우리의 공간이 더 많은 도시와 사람들의 삶과 교감하고 균형점을 만들어주기를 원한다. 그렇기에 우리가 이야기하는 이 에세이 속 아홉 장의 이미지와 내용들은 아직 정렬되지 않은 우리의 경험과 고민에 대한 파편들의 결합이다. 우리가 만들어가고 싶은 건축은, 개인의 관점을 넘어 세상의 삶과 복잡하고 깊숙한 관계 속에서 끊임없이 변화하고 진화하는 열린 방향성을 가지길 원하기에, 이러한 파편들의 연결고리를 찾는 과정 속에서 아직은 결론내지 못한 현재 우리의 단면임을 밝힌다.

설계자동화
Design Automation

기술과 디자인

컴퓨터를 통해 건축설계를 자동화하겠다는 노력은 지속되어왔다. 1971년 조지 스타이니(George Stingy)는 형태 문법(shape grammar)을 발표했다. 형태의 어휘(vocabulary)는 초기 형태와 이후 생성될 형태들을 구성규칙으로 선택하고 수행하는 논리구조를 포함한다. 현재는 케이피에프(KPF), 숍아키텍츠(Shop Architects), 아디타즈(Aditazz) 및 위워크(WeWork)에 인수된 케이스(Case) 등의 기업이 컴퓨터과학과 건축·도시의 접목을 시도하고 있다. 우리는 아이아크건축사사무소의 사내 기술연구팀인 DCG(Design Computation Group)에서 시작하여, 경계없는작업실의 기술연구팀인 Boudless-X, 그리고 스핀오프하여 창업한 스페이스워크(Spacewalk)로 이어져 기술 지향 건축 연구를 지속하고 있다.

정보의 불평등

우리는 다수에게 의미 있는 기술을 만들고자 한다. 대형 프로젝트에서는 기존의 전문가 시스템이 잘 작동한다. 하지만 소규모의 경우에는 기존 방식으로 해결하는 데 한계가 있다. 보통 작은 프로젝트가 대다수이다. 우리는 이 지점을 기술로 해결하고자 한다. 대한민국의 중소형 주택들이 우리의 도전 대상이다. 표준 상품으로 유통되는 아파트는 관련 데이터와 지표들도 잘 축적되어 있고, 단위 규모가 커 전문가가 좋은 방향으로 문제를 해결하기에 유리하다. 16억m²의 전체 주거시설 중 소형 주택은 전체 면적의 40%를 차지하지만 동 개수로는 96%를 차지해 작지만 수가 많다. 하지만 건축 관련 23개의 법 중 건축법만 해도 일 년에 여섯 차례 변하기 때문에 대부분의 사람들은 정보에서 빠져 있다.

모두를 위한 건축

경계없는작업실 고객 중 상당수가 우리가 제안한 최소 설계비를 듣고 고민한다. 맞춤형 고급정장은 그 나름의 매우 훌륭한 가치가 있다. 하지만 SPA 브랜드처럼 모두를 위한 옷도 있어야 한다. 모두를 위한 옷은 기술에 기반해 다품종 소량생산이 가능한 제품을 개발할 때 만들 수 있다. 우리는 기술을 활용하여 전문가 시장에서 소외된 다수를 위한 건축을 지향한다. 정보를 평등하게 제공하는 것은 문제를 해결할 수 있는 첫 단계라고 생각한다. 우리가 개발하는 서비스는 웹이나 모바일로 접속해 누구나 정보를 얻을 수 있다. 또한 데이터와 전문가 자문을 결합한 서비스를 기존 시장에서 접하지 못하던 저렴한 가격으로 제공하는 실험을 하고 있다.

가치평가 모델

우리가 기술자산으로 축적하고 있는 것은 건축 인공지능을 통한 부동산 개발 가치평가 모델이다. 단일 필지나 합필 후 등 다양한 시나리오에서 개발된 뒤의 가치를 측정할 수 있다. 모델을 개선할 때 실제 데이터를 많이 활용한다. 우리는 서울 시내에서 2년 내 신축된 공동주택의 필지를 대상으로 건축 인공지능으로 연산하고, 건축물대장의 건축개요와 연산된 자동설계 건축개요를 비교해 차이가 큰 부분의 원인을 찾고 모델을 정교화하는 작업을 반복한다. 우리는 가치평가 모델을 지속적으로 개선해 부동산 개발에 드는 많은 사회적 비용을 낮추고 토지를 가장 효과적으로 활용할 수 있는 방법을 찾고자 한다.

문주호 임지환 조성현

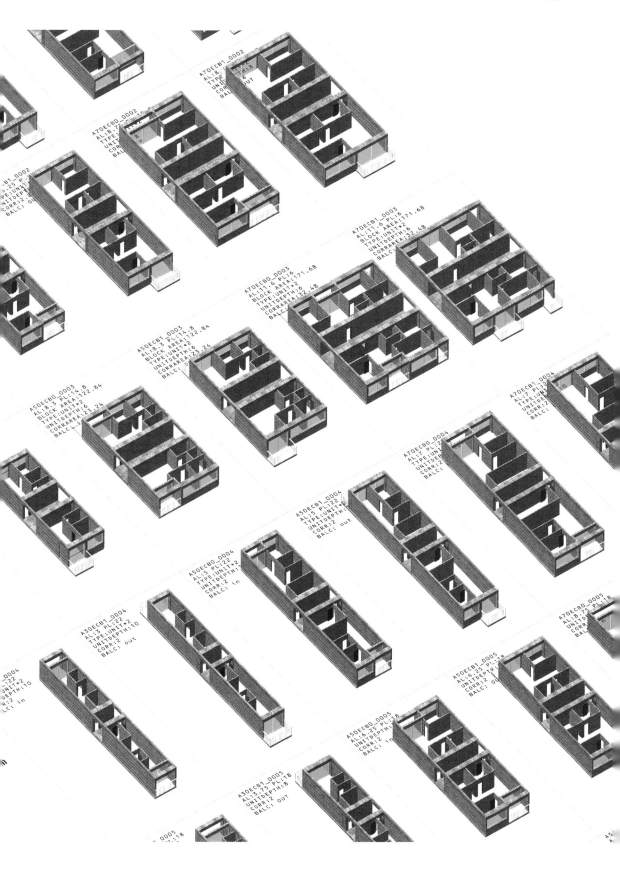

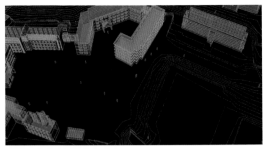

2012 부산대학교 마스터플랜 프로토타입(아이아크)

마스터플랜 프로토타입으로 건축설계 방법을
입력하지 않고 정량적 평가가 높은 오픈스페이스와
건물 배치를 찾을 수 있는지 2일 정도 테스트하였다.

2012 배제대학교 마스터플랜 연구(아이아크)

학생의 시간표, 동선, 시각적 음영, 에너지,
일사 등 가능한 많은 데이터를 넣고 자동화를 시도한
프로젝트로, 구체적인 답을 구하지는 못했다.

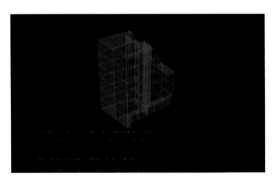

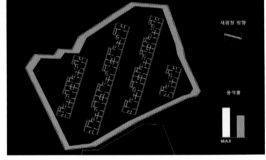

2013 소형 필지 설계자동화 프로토타입

경계없는작업실 초창기에 법규 제약이 많은 소형 필지
작업을 자동화하고자 프로토타입을 만들었다.
소규모 사무실에서 전력이 분산되어 운영이 어려워져
2주 후 중단하고 건축설계에 집중하였다.

2015 가로주택정비사업 설계자동화 연구

사무실 운영이 안정화되면서 기술팀 Boundless-X를
조직하고 가로주택정비사업 설계에 대한 자동화를
시도해 프로토타입을 만들었다.

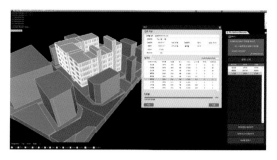

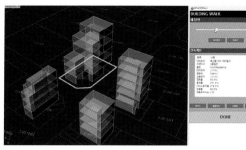

2016 1차 자동화 제품(서울주택도시공사)

앞선 가로주택정비사업 자동화 프로토타입을
바탕으로 1차 제품화하여 서울주택도시공사에
납품했다. 주민을 대상으로 한 신속한 사업성 검토
프로세스에 지금까지 사용되고 있다.

2017 소형 주택 설계자동화 연구

소형 주택에 대한 새로운 설계방법론을 알고리즘화하고
연구버전을 만들었다. 현재 이 연구를 기반으로
많은 사람이 사용할 수 있도록 제품화 중이다.

서울 시내 2년내 소형 건물 신축사례
건축개요와 알고리즘 차이 그래프

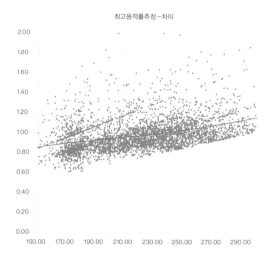

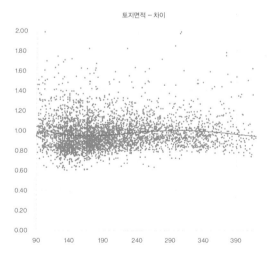

중규모 공동주택

2016 가로주택정비사업 자동화 소프트웨어
초기 기계학습인 유전 알고리즘을 활용하여 대지
위에 용적률이 가장 높은 배치를 만들고 실을 나누고,
데이터베이스에 있는 평면을 늘리는 방식으로
적용하였다. 조사된 시세 등의 값을 입력하여
사업성 검토를 수행한다.

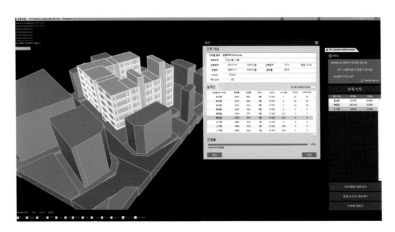

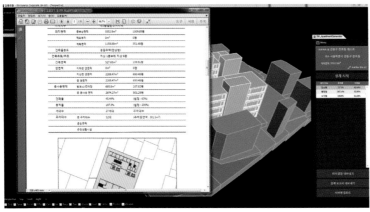

문주호 임지환 조성현

2018 가로주택정비사업 자동화 소프트웨어

입력된 평면 데이터베이스에서 최적의 조합을 찾는 형태로 진화하였다. 수천 개의 평면 조합을 1분 안에 검토한다. 도시 데이터에 기반한 종전 종후 자산을 추정하는 서비스도 테스트 하고 있다.

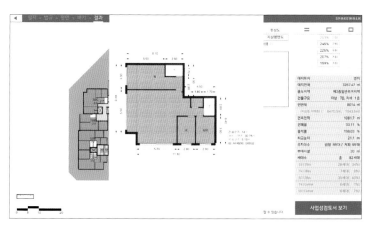

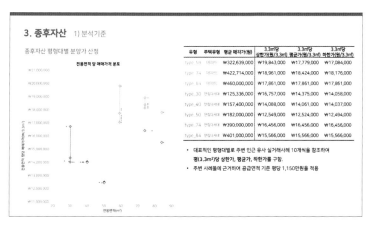

3. 종후자산 1) 분석기준

종후자산 평형대별 분양가 산정

유형	주택유형	평균 매각가(원)	3.3㎡당 상한가(원/3.3㎡)	3.3㎡당 평균가(원/3.3㎡)	3.3㎡당 하한가(원/3.3㎡)
type_59	아파트	₩322,639,000	₩19,843,000	₩17,779,000	₩17,084,000
type_74	아파트	₩422,714,000	₩18,961,000	₩18,424,000	₩18,176,000
type_84	아파트	₩460,000,000	₩17,861,000	₩17,861,000	₩17,861,000
type_30	연립다세대	₩125,336,000	₩16,757,000	₩14,375,000	₩14,058,000
type_40	연립다세대	₩157,400,000	₩14,088,000	₩14,061,000	₩14,037,000
type_50	연립다세대	₩182,000,000	₩12,549,000	₩12,524,000	₩12,494,000
type_74	연립다세대	₩390,000,000	₩16,456,000	₩16,456,000	₩16,456,000
type_84	연립다세대	₩401,000,000	₩15,566,000	₩15,566,000	₩15,566,000

- 대표적인 평형대별로 주변 인근 유사 실거래사례 10개씩을 참조하여 평(3.3㎡)당 상한가, 평균가, 하한가를 구함.
- 주변 사례들에 근거하여 공급면적 기준 평당 1,150만원을 적용

부천시 역곡동 대림아파트

주소 : 부천시 역곡동 104-9
용도지구: 2종일반주거지역
건축용도 : 아파트(리모델링이 용이한 구
건축규모 : 지상13층, 지하1층
용적률 : 245.69%
세대구성 : 59 ㎡ 25평형 78세대
비례율 : 97.56%

시흥시 은행동 철산연립

주소 : 시흥시 은행동 244
용도지구: 2종일반주거지역
건축용도 : 아파트(리모델링이 용이한 구
건축규모 : 지상13층, 지하2층
용적률 : 247.95%
세대구성 : 30 ㎡ 14평형 24세대
          59 ㎡ 24평형 51세대
          총 75세대
비례율 : 100.04%

부천시 원종동 광성/보원아파트

주소 : 부천시 원종동 131
용도지구: 2종일반주거지역
건축용도 : 아파트(리모델링이 용이한 구
건축규모 : 지상12층, 지하2층
용적률 : 249.68%
세대구성 : 40 ㎡ 16평형 105세대
          50 ㎡ 20평형 45세대
          59 ㎡ 24평형 60세대
          총 210세대
비례율 : 80.09%

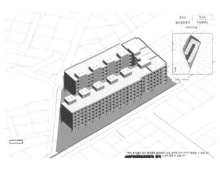

수원시 세류동 태

115세대
104.57%
분담금 1.1억원

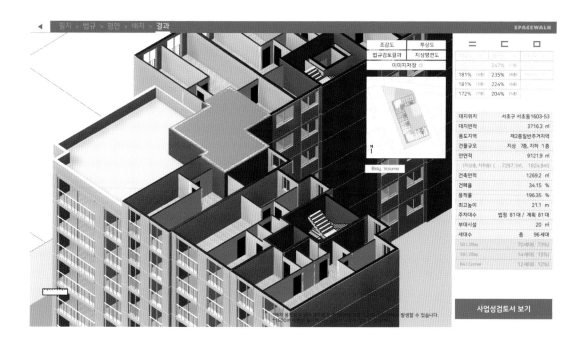

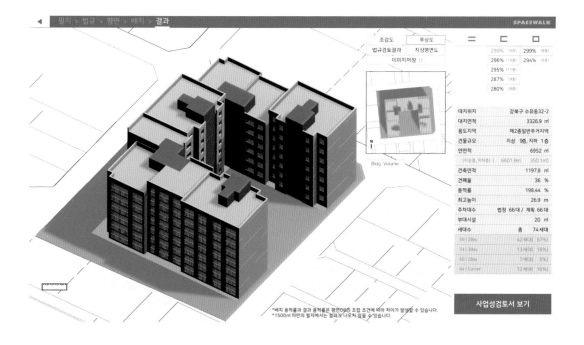

중규모 공동주택(해외)

2017 베트남 사회주택 공급 스마트 설계 프로그램
KOICA CTS 프로그램에 선정되어 베트남의
저소득층을 위한 사회주택 투자검토 소프트웨어의
프로토타입을 만들어 보급했다.

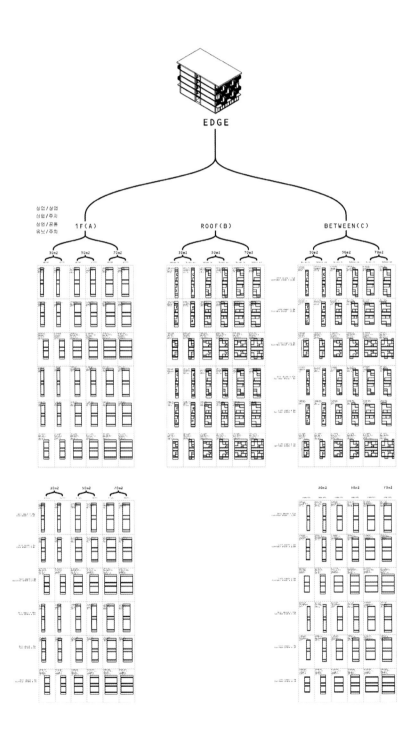

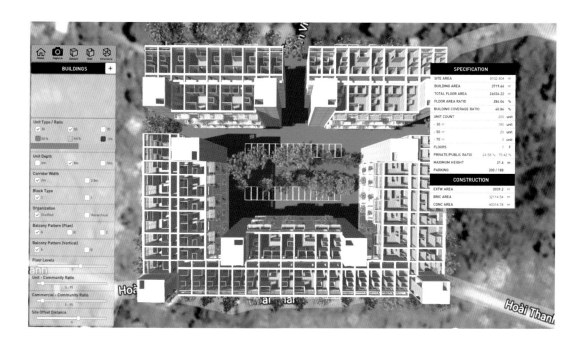

설계자동화

소규모 주택

소규모 주택은 시장에서 가장 방치되어 있는 주택 상품이다. 서울시에만 아파트가 아닌 주택이 98만 호 이상이 있고 이 중 56%가 25년 이상 노후화되었다. 이곳에 사는 많은 시민은 주택을 개발할 역량 및 정보가 없어서 건설업자에게 집을 팔고 이주해야 한다.

2016 소형 주택 연구 01
GIS와 연결해 토지가격이 같다면 어떤 토지가 좋은 자산인지를 판별하는 프로토타입 연구

2016 소형 주택 연구 02
앞선 연구를 발전시켜 층별 개요를 개략적으로 산출할 수 있게 개발

2017 소형 주택 연구 03
주차장 모듈 등을 연결해 소형 필지 설계의 정확도를 향상시킴

2018 소형 주택 설계자동화 제품
주변 도시 데이터와 함께 건축 가능성을 빠르게 판단할 수 있는 제품의 프로토타입

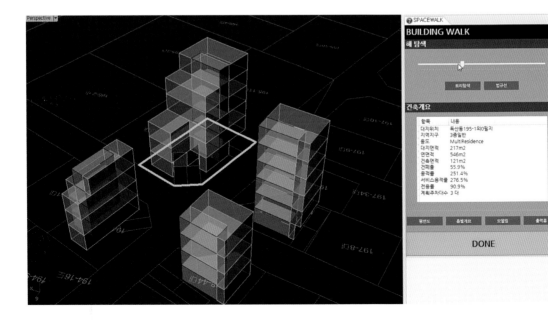

랜드북

랜드북은 데이터와 건축설계 엔진에 기반하여 시민에게
개발 정보를 제공해주는 웹서비스이다. 프로토타입 테스트를
지속하고 있다.
http://www.landbook.net

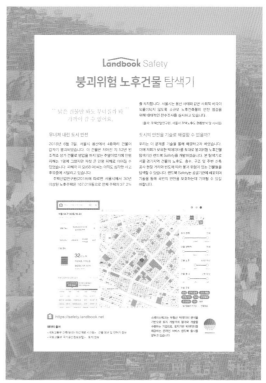

그린램프라이브러리 독서실
Greenlamp Library

학습에 집중하는 공간

그린램프라이브러리는 학생을 대상으로 하는
프리미엄 독서실이다. 경계없는작업실은
그린램프라이브러리 사업의 기획 단계부터 디자인에
참여해, 시각 디자이너와 브랜딩을 협업하며 인테리어
계획을 진행했다. 학생의 공부 행태를 분석해 총
여섯 종류의 좌석을 계획했고, 넓고 쾌적한 라운지를
제공해 공간의 질을 높였다.

사업 기획 단계부터 참여

프리미엄 독서실을 표방한 그린램프라이브러리에
사업 단계부터 참여해 주도적으로 공간 구성을 만든
프로젝트다. 우선 전반적인 브랜드 이미지를 구축하고
이를 바탕으로 인테리어를 계획했다. 브랜드와
공간의 밀접한 연결을 중시하며 통합적인 브랜드
아이덴티티를 구축했다.

열린 도서관 같은 독서실

그린램프라이브러리는 학생을 대상으로 하는 학습
공간이지만, 독서실보다는 도서관으로 쓰이길
바랐다. 고전적인 도서관의 컬러코드, 가구 및 조명을
모티브로 하여 현대적인 방식으로 재해석했다. 그중
오래된 도서관에서 공통으로 발견되는 그린램프를
브랜딩의 시작점으로 삼았다.

학습 행태에 따른 여섯 가지 좌석

그린램프라이브러리에서는 학생이 그날그날 공부
행태에 따라 여섯 가지 선택지를 가지고 좌석을
자유롭게 이동하며 공부할 수 있다. 좌석의 종류는
단순히 가구의 변형보다 공간의 위계와 관계의
개폐를 통해 구분했다. 기존의 독서실은 I이용자
I좌석의 원칙으로 좌석수의 100%만 회원으로 받을
수 있었다. 그린램프는 좌석의 다양화를 통해 공간의
선택지를 넓혔을 뿐만 아니라 자유좌석제로 운영해
사업적으로도 좌석 수 대비 130%의 회원을 받을 수
있는 수익적 이점이 있다.

쾌적하고 넓은 라운지 공간

자유좌석제를 통해 좌석 수 대비 130%의 회원을
확보한다는 것은 단순히 수익성을 높인 사실
그 자체보다 큰 의미를 가진다. 동일 면적에서 좌석
수를 줄여도 수익이 같을 수 있기 때문에 공용 공간을
늘려 쾌적하고 넓은 라운지 공간을 이용자에게
제공할 수 있었다.

IT와 연계한 피드백

물리적인 하드웨어뿐 아니라 소프트웨어적으로도
학습에 집중하는 분위기를 조성하려 했다. 그중
하나로 '시간이 실력이 되는 곳'이라는 슬로건
아래 이용자의 학습 시간을 라운지에 있는 디지털
보드를 통해 보여주고, 이용 패턴 등의 정보를 월간
리포트로 제공했다. 디지털 보드에는 어제 가장 오래
공부한 학생의 이름, 합격자의 평균 공부 시간 등을
게임 정보처럼 제공했고, 이에 건강한 자극을 받아
오랫동안 집중해 공부하는 문화가 형성되었다.
또한 설계자는 좌석별 선호도에 대한 정보를 받아
시간이 지날수록 진화하는 공간을 만들 수 있었고,
지금도 계속 그린램프는 변화하고 있다.

문주호 임지환 조성현

그린램프라이브러리 독서실

그린램프라이브러리 독서실

문주호 임지환 조성현

그린램프라이브러리 독서실

문주호 임지환 조성현

그린램프라이브러리 독서실

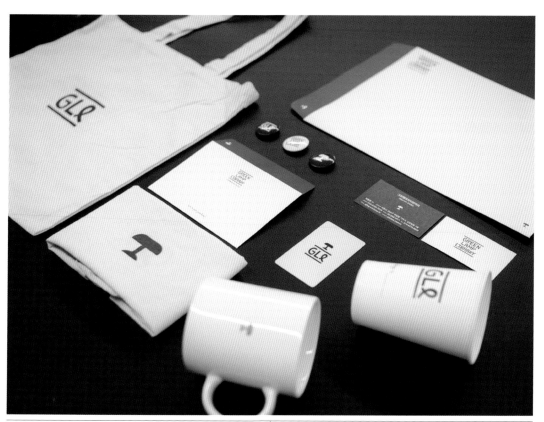

설계	경계없는작업실(문주호 임지환 조성현)
위치	전국 40여 개 지점
용도	독서실
규모	300m² 이상
내부마감	석고보드 위 페인트, 나무
시공	디자인 하울
설계기간	4주
시공기간	5주
사진	신경섭

후암동 복합주거
Huam-dong Multiplex

토지부터 찾기

후암동 복합주거는 은퇴를 준비하는 노부부를 위한
집이다. 건축주는 적막한 교외보다는 생기 있고
자연환경이 나쁘지 않은 소박한 동네를 찾았다.
그리고 노후를 위해 적당한 임대 수익이 가능한
곳에 살기를 원했다. 우리는 조건에 맞는 땅을 찾는
작업부터 시작했다. 후암동 대상지는 위의 조건을
적절히 갖추었고, 이형의 대지 탓에 경쟁력 있는
가격으로 시장에 나와 있었다. 그리고 대상지 후면의
인접 가로에 소월길과 연결된 엘리베이터가 생기면서
유동인구가 늘어, 상가로 발전될 가능성도 있었다.
위치상 서울 시내 조망과 채광을 확보할 수 있어
매입을 신속하게 결정했다.

인접 도로에서의 유입

대상지가 경사지라서 주도로와 2층의 눈높이가
맞는다. 따라서 2층까지 상가로 계획하고, 최대한
가운데를 비워서 답답한 느낌을 없애 사람들이
올라가고 싶은 공간을 만들었다. 또 하나의 과제는
주 가로에서 보이는 건물의 이미지였다. 이쪽에선
북쪽 일조권 높이 제한으로 단차가 생겨 건물의
후면처럼 보일 수 있었다. 우리는 오히려 단을 디자인
언어로 활용하여 전체 조형의 통일감을 확보하고,
후면도 전면처럼 디자인했다.

프로그램 배분

1층은 주차장을 제외하면 면적이 매우 작다.
그래서 1층에 작은 주방과 손님을 맞이할 공간을
배치하고 1층과 지하층을 연결하여 하나의 공간으로
사용하도록 했다. 2층은 중간을 비우고, 양쪽에 각각
상가를 배치했다. 3층에는 11평 남짓 되는 3개의 주거
유닛과 4, 5층에는 각각 하나의 주거 유닛을 두었다.

지어진 후

건물이 지어지고 입주자와 상가 임차인이 들어오면서
주변이 변화하고 있다. 방문자는 SNS에 건물을
배경으로 각자 인상적인 풍경을 담은 사진을 올린다.
골목 방향으로 열려 있는 2층 공용부에 테이블을
두고 앉아 이야기를 나눈다. 지나가는 사람들은
시원하게 뚫린 공용부를 바라보며, 신기한 눈길로
건물을 바라보곤 한다.

후암동 복합주거

후암동 복합주거

문주호 임지환 조성현

후암동 복합주거

후암동 복합주거

문주호 임지환 조성현

매스 스터디
Mass Study

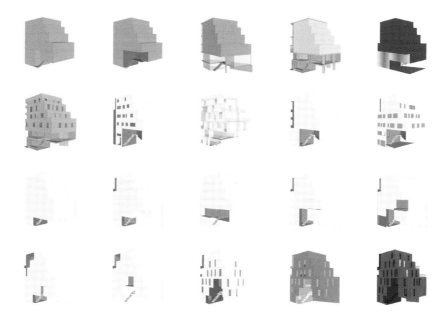

설계	경계없는작업실(문주호, 임지환, 조성현)
설계담당	강가윤
위치	서울시 용산구 후암동 43-5
용도	근린생활 시설, 다가구주택
대지면적	216.78m²
건축면적	129.25m²
연면적	494.36m²
규모	지상 5층, 지하 1층
높이	18.96m
주차	5대
건폐율	59.62%
용적률	199.48%
구조	철근콘크리트조
외부마감	점토벽돌, 갈바륨 위 도장
내부마감	석고보드 위 페인트, 나무
구조설계	터구조
기계설계	타임테크
전기설계	극동파워테크
시공	태인건설
설계기간	2015. 10 – 2016. 7
시공기간	2016. 8 – 2017. 8
사진	신경섭

입면도
Elevation

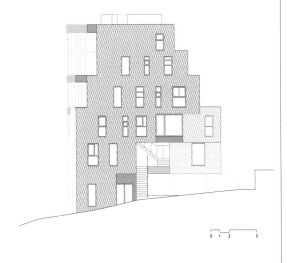

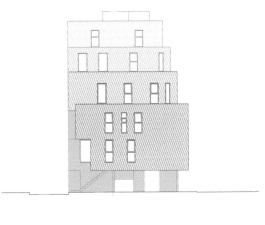

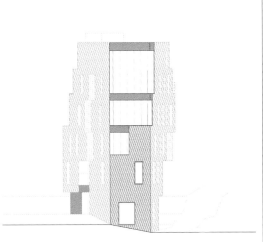

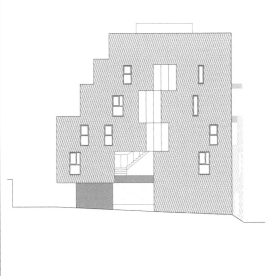

평면도
Plan

3F

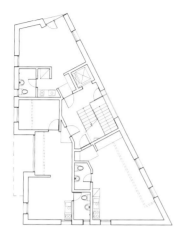

4F

1F

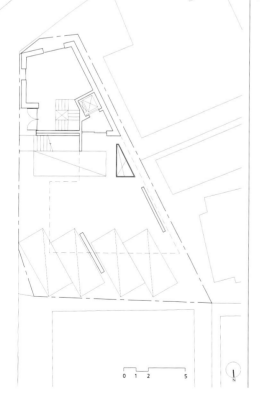

0 1 2   5

N

2F

논현동 코너하우스
Corner House

작은 대지 속 다양한 요구 수용하기

저층부에 상업시설이 가능한 대지는 I층이 분주하다. I층은 물론이고 2층 상업시설과 지하층도 별도의 출입 동선을 요하며, 상부의 주거는 프라이버시를 위해 분리된 동선과 분위기를 원한다. 거기에 법적 주차시설도 필요한데, 모든 요구 조건을 만족하기란 쉽지가 않다. 40평의 작은 대지에 이 모든 것을 수용하고 디자인 완성도 놓치지 않기 위해 치열한 고민에 들어갔다. 대지의 작은 수직적 레벨 차이를 활용하는 것부터가 해결의 시작이었다.

레벨 차이를 이용한 계획

한정된 면적 안에서 요구 조건이 많아 평면만으로는 해결할 수 없었고, 대지의 레벨 차이를 이용해 해결해나갔다. 별도의 출입구가 필요한 시설은 반 층 레벨 차이를 두어 동선을 분리했다. 주거의 출입구는 건물의 후면으로 배치하여 사생활 보호와 동시에 상업시설의 전면성을 함께 확보할 수 있었다. 지하로 내려가는 출입구는 코너의 급경사지를 이용해 높은 층고를 주어 지하임에도 쾌적하게 느껴진다.

수익성을 고려한 디자인

똑같은 면적이라도 프로그램과 레벨에 따라 임대료가 달라진다. 특히 I층 임대료가 가장 비싼데, 주차 때문에 I층 상업 공간 면적을 많이 확보하기란 쉽지 않다. 코너하우스에서는 경사를 이용해 지하 출입구에 높은 층고를 확보해 임대료를 20% 올렸고 I, 2층을 스킵플로(Skipflow)로 연계해 더 큰 수익을 불러왔다.

인장 기둥으로 1층 주차 공간 확보

각 시설이 요구하는 4개의 진입구를 모두 해결하니 주차 공간이 빠듯했다. I층 근린생활 시설 진입구와 지하 I층 진입구 사이를 주차 영역으로 만들자 구조 기둥을 위한 자리조차 남아 있지 않았다. 기둥을 땅으로 내리는 대신에 인장부재를 넣어 주차 공간을 확보할 수 있었다. 힘은 인장재를 통해 다른 기둥으로 옮겨졌다.

다양한 방향과 레벨을 고려한 입면

오거리 코너에 위치한 코너하우스는 여러 각도와 레벨에서 보인다. 특히 건물 정면에 위치한 경사지에서는 법적 제한선을 활용한 오각형 지붕까지 한눈에 읽히기 때문에 외벽 재료와 함께 지붕 재료도 어둡지만 반짝이는 석재로 마감했다. 석재 고유의 반짝임은 빛의 방향에 따라 다양한 색으로 변한다. 반면 저층부의 상업시설을 상층부와 대비되는 투명한 유리로 마감했다.

밝고 재미있는 주거 공간

개방된 저층부의 상업시설과는 달리 상부의 주거 부분에서는 프라이버시를 지켜야 했다. 또 거주성을 위해 채광도 좋아야 했다. 이런 기본적인 사항과 더불어 각기 다른 용도의 공간이 한 집 안에서 어우러지는 흥미로운 공간을 만들고자 했다. 외부에서는 닫혀 보이지만 천창과 남측의 채광창이 밝은 공용 공간을 제공해 기분 좋은 진입 공간을 만들었고, 수직 조닝을 통해 작지만 다양한 공간을 품는 집을 완성했다.

문주호 임지환 조성현

논현동 코너하우스

문주호 임지환 조성현

논현동 코너하우스

문주호 임지환 조성현

논현동 코너하우스

문주호 임지환 조성현

논현동 코너하우스

논현동 코너하우스

설계	경계없는작업실(문주호, 임지환, 조성현)
설계담당	심규대, 강경국
위치	서울시 강남구 논현동 188-23
용도	근린생활 시설, 다가구주택
대지면적	132.3m²
건축면적	79.37m²
연면적	348.12m²
규모	지상 5층, 지하 1층
높이	19.8m
주차	3대
건폐율	59.99%
용적률	199.86%
구조	철근콘크리트조
외부마감	지정석재, 갈바륨 위 도장
내부마감	석고보드 위 페인트, 나무
구조설계	터구조
기계설계	정인엔지니어링
전기설계	대경ENC
시공	태인건설
설계기간	2013. 11 – 2014. 4
시공기간	2014. 5 – 2015. 1
사진	신경섭

단면도
Section

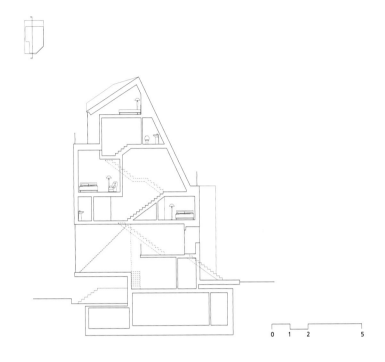

0　1　2　　　　5

평면도
Plan

0　1　2　　　　5

B1

1F

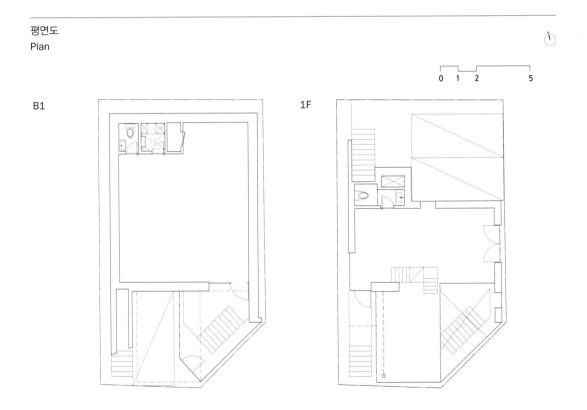

3MF

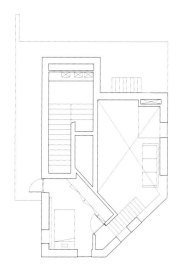

5F

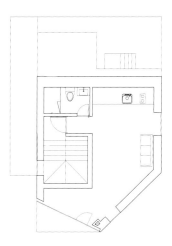

2F

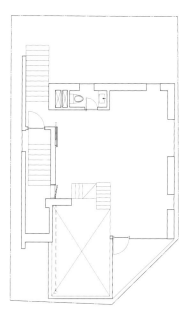

3F

비평

경계없는작업실에게

이민아 · 건축사사무소 협동원 대표

이 글

어찌 보면 나는 세 사람의 글과 말을 비평하는 셈이다. 상을 지원하며
제출한 작품 몇 개의 개별 크리틱이 필요하다는 생각은 하지 않았다. 15년
전 처음 만난 세 사람을 세세히 기억하고 다시 집중하고 그들이 봐 달라고
보내준 많은 새로운 신호를 눈치 있게 읽어냈다. 우리는 오래간만에 긴
이야기를 나눴다. 수상자의 멋쩍은 교만이 내게 들킬 줄 알았는데 사나운
건축판에 빨리 나와버린 조숙아의 불균형감과 애늙은이의 피로 상태
같은 것이 동시에 느껴졌다. 그러나 그 속에는 놀라운 깊이의 사유와
'이제 어느 방향으로 어떻게 갈 것인가'의 쇠못 같은 질문과 자신들 작업의
근본을 묻는 철저함이 있었다. 상 받는 것이 그렇게 중요하더냐를 물을까
하다가 못했다. 사실 몇 번 젊은건축가상 심사를 하며 언젠가 이 셋이
나타나기를 내심 바라기도 했다. 이 글은 경계없는작업실, 세 사람이
읽어주는 것만 염두에 두고 썼다. 글을 쓰며 나의 건축 작업에 대해서도
많은 생각을 했다.

1학년

언제부터인가 셋이 일을 같이 한다는 소식을 접했을 때 문주호(이하
문)와 조성현(이하 조)만큼 임지환(이하 임)에 대한 기억이 바로 떠오르지
않았다. 셋이 뭉쳤다는 소식은 셋의 행보를 SNS로 노려보고 있던 열성
후배들이 뭉치기 조짐이 보일 때부터 빅뉴스라며 알려줬지만 아주
놀랍지는 않았다. 나는 2003년 1학기에 임을, 2학기에 문과 조를 만났다.
1학년 1학기 스튜디오 배정은 학번 순으로 자동 정해졌고 임은 1학기부터

나의 스튜디오 학생이었다. 건축학과 1학년은 첫 설계 수업의 특성상 이제 서울대생이 되었다는 귀여운 오만감의 충만으로 대부분 설계 강사에 대해 대결의 눈빛을 띠며 수업에 들어온다. 건축은 무엇일까의 호기심 반, 수업 방식이 고등학교 때와는 굉장히 다르지만 뭐 나는 문제 없음의 자신감 반으로 학기가 시작되고 선생과 가벼운 신경전을 벌인다. 임도 그런 편이었다는 것을 2학기 문과 조를 떠올리며 기억해냈다.

2학기 스튜디오 배정은 학생의 선택이었다. 1학기 때 옆 스튜디오의 특이한 인물로 유심히 봤던 조와 문이 내 반에 왔다. 건축학과 1학년 2학기는 나름 위기가 감도는 시기다. 상상치 못했던 수면부족, 만들기에 대한 스트레스를 경험한 첫 학기와 그 파장으로 소중한 첫 방학을 서툴게 보낸 직후다. 귀여움은 사라지고 전공에 대한 자신의 입장을 스스로에게 지독히 설명해봤을 법한 표정들로 선생을 마주한다. 그 와중에 문의 몰입력은 독보적이었다. 선생에게 보여주는 저 많은 희한한 스케치들조차 더 많은 스케치들 가운데 골라온 듯했고, 생각이 요령 있게 설명되지 못할 때는 처음으로 돌아가 반복하는 집요함으로 뒷 순서 학생의 시간을 빼앗기 일쑤였다. 우리의 대화는 튜터링의 범주를 벗어나 끝없이 멀어지기도 했으나 그 기억이 강렬하다. 조는 포항공대를 다니다가 온 동네 형 같은 학생이었다. 과제를 때맞추어 잘 해오는 편은 아니었다. 대신에 저렇게도 할 얘기가 많을까 싶을 정도의 확신에 찬 동문서답이 선생을 자극했다. 학생들은 조와 나의 수업을 엿들으며 설계 과제의 방식에 갇혀 있던 자신들을 위로하고 조의 능청스럽고도 유연한 두뇌를 신기해했다. 이들과는 달리 1학기의 임은 예리했지만 조용했다. 날카로운 질문도 고의로 투박하게 했던 것 같다. 교감을 즉각적으로 표출하거나 선생에게 응석부리는 학생들과는 달랐다. 자신의 속도와 에너지를 조절할 줄 아는 영민함으로 수업의 흐름에는 끌려가기 싫어했다. 세 사람의 공통점은 약간 어중간한 반골 기질이었던 것으로 기억한다.

그리고 몇 년 전, 경계없는작업실이 잘나간다며 열성 후배들이 보여준 유명 잡지 사진에서, 턱수염을 기른 문주호와 기름기 도는 반짝이는 조성현의 얼굴을 보았다. 그 가운데 변함없이 소탈하게 미소만 짓는 임지환의 롤이 오히려 궁금했고 문과 조의 사장님 이미지를 진심으로 염려했다.

사무실

세 사람이 팀을 만들었다는 것에 대해 나는 몇 가지의 선입견을 가지고 있었다. 이 셋에게는 서로를 진정으로 잘 교환하는 힘이 있을까?

사무소 이름은 왜 이렇게 정했을까? 경계의 포괄성으로 건축의 본질을
얘기하고 싶은가? 넘나들며 무슨 일이든 다 한다는 영업 전략을 드러내는
건가? 혹은 경계는 사라져야 한다는 선언을 담고 싶은가? 한번 들으면
기억되는 이름을 짓고 싶었나? 그러나 '무엇 없는 무엇'은 꽤나 구식의
개념 규정으로 셋의 합의로 골랐다고 보기에는 어쩐지 진부했다. 게다가
서울대학교 출신자의 팀이라는 마케팅 속이 뻔히 드러나는 것에 대한
실망도 컸다. 그러면서도 현실과의 독한 대면을 유예하지 않고 어느덧
준공작을 발표하기 시작하는 이 셋이 좀 궁금했다.

세 사람을 만나러 가는 큰 목적은 인터뷰보다 사무실 구경이었다.
성수동 건물의 한 층, 한 켠에 자리 잡은 모습은 의외였다. 마치 움직이는
열차의 자리 하나씩을 임시로 점유하듯, 물리적 공간쯤 중요치 않음에
산뜻하게 도전하고 있었다. 자신만의 고정 공간에 대한 집착을 내려놓고,
내가 다가가니 문은 가볍게 일어나 열차의 식당 칸으로 잠시 옮겨가듯
나를 끌고 유유히 작은 회의실로 이동한다. 사무실 내 자리. 나를 둘러싼
물건들의 무게, 나를 구속하는 기록들의 누적, 바닥이 꺼질 듯 쌓아올린
책들의 풍경에 집착과 미련을 두던 나는 이 경계 없음에 위축되었다.
비장소적인 낯섦 속에서 내가 미개인 같았다. 이들이 각성하는 무의미한
경계가 상투적 공간을 재정의하고, 옆 칸에 앉은 다른 분야 전문가와의
시너지를 도발하고 있었다. 경계없는작업실의 현재적 정체성을 이해하는
데 도움이 되었다.

경계없는작업실은 2013년 만들어져 3년이 되던 2016년 여름, 자발적
위기를 마주한다. 매출도 늘고 현장 수도 늘고 직원 수도 늘었는데
셋은 스스로 행복한가를 질문한다. 그리고 감히 세상을 이롭게 했는가를
묻는다. 셋은 동업을 계획할 때 이미 이 질문을 예비했을 것이다.
2009년 팀을 만들며 오래 못가 망하더라도 선택할 수 있는 다음 카드는
여전히 많다고 무장한다. 성장통 속에서도 '그래서 우리는 무엇에
헌신하고 있는가'의 질문을 거듭하며 이 성찰이 어떻게 실천적 덕목으로
사회와 건축주와 건축가에게 구실할 수 있는지 서로에게 지독하게 묻기
시작한다. 전략을 정비하려는 노력과 동시에 셋은 개별적인 성장을
시도한다. 1년여 후 셋의 열성 후배들을 통해 나는 경계없는작업실의
해체 소식을 접한다. 경계없는작업실은 후배들의 본보기이자 취직을
희망하는 회사 1순위였다. 해체 소식은 경계없는작업실의 상징적 역할의
상실에 대한 허탈감을 주면서도 진정 모든 경계로부터 초월했다면
가능한 결정이었을 것이라는 짐작도 가능케 했다. 셋은 변함없이 분주한
가운데 짧은 좌절과 검증을 자청했을 것이다. 그리고 셋은 초월과 천착을
반복하며 서로에게 낱낱이 내보여준 욕망과 능력으로 각자의 바닥을
더 굳혀 다시 모였다.

글, 말 ————————————————————————

셋은 많은 종류와 양의 텍스트를 보내왔다. 나는 셋이 자신의 문장
속에서 주로 사용하는 단어들을 구분하며 셋이 글과 말을 통해 가까이
붙잡고 싶어 하는 개념들에 주목했다. 셋의 언어에는 다듬어지지 않은
공통적인 따분함이 있었지만 사회 변화에 대한 정확한 감지와 문제의식
높은 통찰도 읽혔다. 이들의 글쓰기는 언어적 유희를 허용하지 않겠다는
건조한 서술로 일관한다.

문의 에세이는 설계 과정 속에서 혹여 자신이 궁극적으로 탐구하기를
갈망하는 공간의 본질에 대해 잠시 잊을지도 모른다는 염려로 가득
차 있다. 그래서 자석으로 쏠려가는 쇠붙이처럼, 개발 이익, 수익성
등을 언급하다가도 곧장 돌아와 좋은 공간의 추구와 건축가의 직능을
강조한다. 셋은 기성 건축가가 예민하게 고르는 감각적 문체를 모방하지
않고 현실 속 숙제로 보이는 단어만 조합하지만 이들의 서툰 글 솜씨는
오히려 미래적이다. 그 와중에 문은 '우리도 건축가니까 비우는 것을
좋아한다'는 말을 흘린다. 결국 건축가는 자신이 한 일에 대해 글을 쓰고
건축가가 선택하는 어휘의 비약은 그들의 고뇌가 어느 지점인지를
드러낸다. 건축적 글쓰기는 매우 위선적일 수 있는 도구이면서
유일하게 순간마다 자신을 성찰하게 만드는 고해 과정이기도 하다.
경계없는작업실의 텍스트는 현학적인 어휘를 마다하고 자신에게
당장 절실한 단어만 반복해 못 박고 있다. 개념어를 가공해 창작의
동력원으로 삼아 지적 반향을 불러일으키고자 했던 선배들의 허식과는
다르다. 산만한 경험과 독서가 제공하는 단편들을 동원해 그것을
건축적 글쓰기로 여겨온 나로서는 군살 없이 골격만 버티고 있는
셋의 재미없는 글이 꽤 신경 쓰인다.

조의 말투에는 자신이 하는 일을 노련하게 설명하기 위해 개발된 기계음
같은 톤이 있다. 마치 자신이 구사하는 의미와 음성마저 프로그래밍 하듯
최적화해 두었나 싶은 얄미운 언변이다. 그러나 인도네시아 농민 토지
문제해결을 위한 농촌 마스터플랜 자동화 계획 이야기를 꺼내면서
조의 음성은 문득 가늘어지며 흔들렸다. 조를 떨리게 하는 것은 무엇인가?
선한 일에, 공적 가치가 뚜렷한 일에 자신을 잡아다 놓고 싶은 저 속을
내 앞에서 뭐라 설명할 길이 없었겠거늘 설득과 고백의 언어가 조도
모르는 사이에 진정성 있게 교차하고 있었다.

임이 뒤늦게 보내준 텍스트까지 읽으며, 마침내 나의 문제를 마주했다.
삼십 대의 나는 무엇을 건축이라 믿었는가? 임지환의 시간을 나는
어떻게 살았는가?

작업, 작업 너머 ————————————————————

준공작을 처음 본 것은 몇 해 전 월간 《공간(SPACE)》 피어리뷰에서였다.
블라인드 평가임에도 셋의 작업인 줄 즉각 알았고, 셋이 각자 수련했던
설계사무소 소장님 세 분도 오버랩 됐다. 나는 셋의 작업을 선정에서
제외했다. 개발 이익, 합리적 공간 활용, 조형성이라는 세 가지 성취에
대한 설계 의도가 잘 읽혔음에도 이 건물이 셋의 작업이라는 추측
때문에 블라인드 평가를 역이용한 주관적 결정을 했다. 셋이 끝까지
가지 않았음마저 읽혔기 때문이다. 설계 진행 과정에서 공간의 본질에
대해 격론을 벌였더라면 또는 서로의 감수성과 철저하게 싸웠더라면
했다. 이 조형은 도시에 적절한가, 왜 이 재료이어야 하는가의 감각에
기초한 질문까지도 이어졌어야 했다. 이쯤에서 멈춘 것은 셋이어서
그래야 했던 타협인가, 그냥 디테일을 잘 몰라서인가의 복잡한 의구심이
리뷰를 괴롭혔다. 실은 매번 치열한 공방과 설득의 시간을 거친다는
것을 인터뷰를 통해 알게 되었다. 셋은 룰을 만들었다. 둘이 반대하면
다시 생각하고 한 명이 강하게 원하면 둘이 다시 생각하기. 셋은 서로를
검열하지만 무릇 건축 공간이 늘 선택의 안전 룰에 따라 걸러질 수
있을까? 혼자 내려놓는 무거운 결정으로 건축의 기본에 가장 근접하게
도달해 있을 때가 어쩌면 더 많을 것이다. 셋은 제자이자 후배로서 역차별
당한 셈이고 나의 결정은 《공간》 피어리뷰 제도의 취지를 비껴갔다.
그러나 건축가가 잡지에 작품을 싣는 것을 셋이 그다지 중요치 않게
생각하기를 진심으로 바랐다.

초기작 도시형 생활주택에서 현재 공사 중인 식물관까지 틀어쥐고
있는 셋의 이른바 지성주의는 외부로부터의 팀에 대한 신뢰를
지속시키는 중요한 무기임에 틀림없다. 그러나 서울대 마케팅으로
시작하여 확고해진 엘리트 오피스의 저력이 항상 무언가 다르게
보여주어야 한다는 자기 속박으로 이전하고, 이로부터 자유로워지지
못한 채 프로젝트마다 점차 과잉의 말과 글이 붙게 된다. 나는 셋의
준공작들로부터 동어반복도 보고 싶지 않았고 설명이 반드시 필요한
특이한 차이가 보고 싶은 것도 아니었다. 셋의 건축에는 단정한 첫인상도
있고 고유의 질감도 있다. 예민한 프로포션도 있고 영리한 디테일도
있고 따뜻한 유머도 있다. 내가 질투를 느끼고 마는 사물과 풍경도 있다.
우리는 왜 이런 이야기를 전혀 하지 않았을까?

논현동 코너하우스와 후암동 복합주거는 이제 셋의 대표작이 되었지만
여기서 강조되는 토지의 가치, 공간의 활용, 공공성 확보에 대해
다른 어느 건축가인들 치열하게 고민하지 않았겠냐고 묻게 된다.
셋의 텍스트에 자주 등장하는 핵심어, '창의적' '극대화' '가치 부여'가

경계없는작업실만의 설계원칙은 아닐 것이다. 그런 줄 알면서도 이 기본적인 것에 더 분투하는 이들이 스스로를 몰아붙이는 저 끝, 헌신의 상대는 무엇인가?

셋은 독보적 아이디어와 지적 감수성으로 독서실이라는 극히 현실적인, 시대의 공간 부산물을 창의적으로 공략한다. 흥미로운 인류학적 그래프와 그래픽을 추출하여 부모와 학생의 욕구를 사로잡고, 못 보던 방식으로 격려해주고, 최적의 공간으로 답한다. 어디까지가 상품이고 어디까지가 세상에의 기여인지 구분은 어렵지만 나는 이제껏 만나보지 못한 독서실의 경이로운 진화에, 이렇게 바뀔 수 있는 것이 또 번쇄한 도시에 얼마든지 있겠다는 기대로 그린램프라이브러리의 공익 버전도 감히 보고 싶어졌다. 만약 폐품 수집센터의 공간과 폐휴지 가격을 체계화하고 좋은 디자인의 야광 셔츠와 방한복을 제공하며, 저소득층 노인에게 자원 리사이클링 최전선에 앞장서는 역할을 정중히 부여하고, 느리게 걷는 노인의 손수레 옆면을 시내버스 광고보다 강렬하고 아름다운 후원 그래픽으로 설계하는 일, 경계없는작업실과 파트너 팀(아토스터디 atostudy)이 그린램프라이브러리의 클린(clean) 버전으로 나서준다면, 셋이 내게 참 거창하게 던졌다고 생각했던 '인간 존엄성의 수호'를 따뜻하게 이룰 것이다.

컴퓨터 능력자 조는 설계자동화를 통해 자동으로 해도 되는 것과 안 되는 것의 경계를 다루며 자신의 기술은 가치판단이 덜 필요한 법규 검토까지만 써먹는 기술이라고 피력한다. 그리고 전문가 시장에서 소외된 보통의 많은 사람이 집 지어볼 엄두가 나도록 정보와 소프트웨어를 공유한다. 그러나 나는 정작 자동화는 신기하지 않고 이 문제를 통해 건축의 본질이 무엇인가의 질문에 재차 붙들리도록 장치를 걸어둔 조가 신기하다. 문과 임에게 설계자동화가 어떻게 체감되는가를 가장 먼저 물었을 테고 셋은 '건축가는 무엇을 하는 사람인가', 결국 또 이 고역의 질문의 시간으로 하루를 결산했을 것이다.

어떻게 계속할 것인가

셋에게서 사장님 미소의 우울은 사라졌다. 셋은 현재의 방식으로 탐구할 것들이 아직 많아 유학을 고민해본 적도, 공공건축가로 선정되고자 지원서를 들고 기웃거려본 적도 없다. 대신 가족을 만들었고 세상의 덧없는 것들과도, 자기 안의 속물스러운 근성과도 한두 번씩 부딪쳐봤을 것이다. 전문가의 허세와 업자의 번민, 열패감도 경험했을 것이다. 그리고 가장 사소한 건축적 실천도 세상살이의 실제적 성장으로서의 가치를 가진다는 것을 목격했을 것이다.

그래, 경계는 없다 치더라도 우리를 절망에 빠트리는 어떤 임계는 엄연히 존재한다는 사실도 깨달았을 것이다. 공간과 장소의 문제에는 의외성이 항상 존재하며 건축과 도시에는 기획이 안 되는 뿌연 영역이 더 넓다는 것도 경험했을 것이다.

우리가 열광하는 건축, 그것은 결국 물질이며 그것에는 물성과 질감이 존재하기에, 직관과 심성에 기초한 결정을 더러 해도 괜찮다고 자신을 설득해보기도 했을 것이다. 건축의 엄결성을 어디서 누구에게 배웠는지 몰라도 내 속에서 허둥대며 건축의 윤리와 대결했던 기억도 있을 것이다. 어느 범주에도 속하지 않는 자신만의 사유의 질서를 건설해보고도 싶었을 것이다. 누구에게나 시간은 교집합적이나 동치일 수는 없으니 셋이 묶여 있는 자리에서 그냥 혼자 빠져버리고 싶은 적도 물론 있었을 것이다. 상을 받고 세상의 활자에 민감해졌을 수도 있을 것이다. 이참에 SNS와 인터뷰를 의식해 스스로를 재정의해보고도 싶었을 것이다. 개발업자의 탈을 쓴 재능기부자가 되고 싶다는 꿈을 진심을 다해 제대로 이야기해보고도 싶었을 것이다. 그러나 건축에서의 봉사는 언어화하는 순간 아무것도 아닌 게 되어버린다는 것도 인식하고 있을 것이다.

건축학과를 수석 졸업했다던 문주호와, 졸업 작품전에서 대상을 받았다던 임지환과, 도널드 트럼프 책을 거의 다 읽었다던 조성현. 건축을 시작하던 시절의 최초의 정열을 기억하는가? 셋은 지금 어느 지점에 있는가?

이민아는 서울대학교 건축학과를 졸업하고 동 대학원에서 석사 학위를 받았으며 베를라헤 건축대학원에서 석사 학위를 취득한 바 있다. 공간연구소, 민현식 건축연구소, 건축사무소 기오헌을 거쳐 현재 자신의 건축사사무소인 협동원을 운영하고 있다. 주요 작품으로 서울대학교 언어교육원, 대전대학교 융합과학관, LH 강남 보금자리 주택, 파주 대교 어린이공연장 등이 있다.

ESSAY

Finding the Boundary

by Jooho Moon, Jihwan Lim, Sunghyeon Cho

1

The dream that I had in my youth must have been based on my interest and curiosity towards buildings. As an architect, I now evaluate buildings from various contexts, but 20 years ago, I simply admired the tallest buildings such as the 63 Building in Seoul and the Sears Tower in Chicago. Instead of Gaudi, on whom I had to study overnight in preparation for my university entrance exam, I was more curious about the heights of the Burj Khalifa that was about to be built in Dubai, the Tokyo Metropolitan Government Building, and the 63 Building. I and my friends compared the number of floors and households in apartments, counted the number of balcony bays on the facades, and admired their grand and wide size. To us of that time, the value of architecture was a result of technology, and a symbol of wealth and growth.

However, despite being the tallest building in Korea, the recently-built Lotte World Tower is interesting but not impressive. The public also do not feel as thrilled from the building's size and height as before. They rather discuss the design of Seoul City Hall and DDP, and the well-designed spaces become designated as hot places as they get introduced and experienced by people. Although the following investigations on spaces are our narratives, but they are all connected to everyone's experience and perspective. Eventually, we are all living and contemplating together as a member of this era.

2

Through diagrams, images, floor plans, and excerpts, we come to explain and persuade to others the intention and the meaning of a building.

We come to decide the direction of a building through numerous discussions with the client, and come to finish a building after a process of arguments, compromises, and persuasions between teammates. Then we ask one another, "is this building beautiful?" It is known that the beauty of a building and its spatiality do not lie in the realm of logics but in the realm of preference. Hence, even something that we had created after spending numerous nights contemplating and have pride in can just simply become meaningless through a single expression of disapproval by the client.

Still, we do not predicate that the beauty of a building depends on preference. It is a cliched example, but the Church of Light by Tadao Ando planted in us an instinctual impression as we began our college studies in architecture. The scenery behind Mont Juic that we coincidentally came across in our Barcelona trip is something that we wish to eagerly recommend to others. The scene of the trees and the sky that come into view as we lay down to rest after a walk in the forest just evoke feelings of happiness. There are tourist attractions that are all about appearances or a certain humanistic context, but places that remain loved by visitors continue to create bonds and share feelings with people.

This is why we continue to challenge the necessities that are expected of space. We question the essential value of space, converse endlessly among ourselves on the perspective of the value-experiencing subject, and try to collect all these as our architectural asset. To put it simply, a house should be rainproof, it also needs to receive ample sunshine and be warm. We strive to be architects with the necessary skills to create deserved things before we go about presenting our perspectives.

3

However, we think that architecture is something that goes beyond mere instinctual value. It is a consequence of attempts to find the balance between a variety of spatiotemporal relationships of the era and the urban agglomeration. Our contemplations should not only be limited to spatiality and the aims of a client. We should both

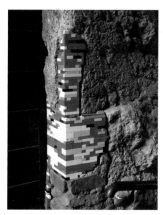

reading the sedimented relationships of the earth that has functioned as the building's borders since long past and also speak about the elements of this new relationship that will now begin to persist after the building's construction.

These are grandiose claims, but ultimately our task is to propose spatial solutions that come from our contemplations on the relationship between road and building, the formational harmony with the surroundings, the appropriate position of the windows, the profitability and utility of the building, and the influences on the neighboring buildings. With the help from technology and imagination, we try to consider as many aspects as possible, and challenge ourselves to bring them to results. In Jan Vormann's work, the deteriorated and altered gaps of a building are filled up with a combination of numerous Lego blocks to almost perfectly restore its outer border. This resembles an spatial analysis diagram such as OMA's Yokohama Masterplan. Each Lego block represents the various themes that architects should reflect onto space, such that the overall appearance is to be composed not out of a single individual's motive but out of an integration of diverse elements and boundaries of contexts. The architectural direction that we are pursuing is not too far from this. Boundless rejects the sense of uniformity that arises from the building's exterior. By endlessly expanding its horizons, it strives to think beyond the fixed ideas to create diverse forms for each site.

4

It is not easy to differentiate a hobby from a specialty, but if we were to understand hobby as something that we can enjoy doing purely for its sake, then our hobby would be map reading. As we carefully observe places in Korea and elsewhere through a map, we often discover things that we have taken for granted and things that are new. Like the color of a traditional roof tile, the color of our scenery in Korea is achromatic and dark blue. Similarly, when zoomed out in the satellite map, the color of the city – whether with roof-tiles or not – and the color of the forest exudes a similar achromatic, navy blue color. Spain or Portugal, when observed from afar, has the color of a red soil, and observing from the urban scale, we realize that the red comes from the houses with red roof tiles and other materials that constitute the urban scenery.

As a basin formed narrowly at the central mountain ridges between Sichuan province (which is the foundation of the state Shu) and Chang'an city, the city of Hanzhong, which is described as a 'chicken rib' (a metaphor for something that is difficult to discard despite its small value) in the Romance of the Three Kingdoms, while acting as an important strategic path between the two places, has a size that is appropriate to its name when compared with the two cities. Cheonan has a good geographical location that is close to both Honam and Gyeongnam. Some cities have a geography created from the collision of tectonic plates due to the volcanic activities in that area.

In this way, by observing a satellite map, one can traverse across hundreds if not tens of thousands of years in time. Over that long period, countless number of people lived their lives and built their cities in that massive geographical space. Not as independent entities, we too become added to this narrative where cities of unmeasurable sizes are being built as a part of the city's continuous time flow. While our architectural activity may seem like an addition of just one more building to the many, however, it is also an activity in which we connect our timeline with this vast tradition.

5
Similar to map reading, urban space exploration also provides us with unexpected encounters and joy. During our travels, we divide the days per city to spend an entire day walking about aimlessly in a certain region. Perhaps it may not be the most effective method, but there is no other method better than this when it comes to travelling and observing a city. Through this, one can also come to experience raw things of the city – things that are not covered in travel magazines or in the map. It gives a pleasing feeling as we discover works by unknown architects, and the vibrant image of the

greatest reason why we are charmed by the port scenery is paradoxically because of the machine's sense of weight and the lightness of its movements. Building-sized containers and ships move about according to their destinations and schedules and diversify the shoreline scenery. The various colors and quantities of the containers create a sense of rhythm, and the movements of the ships and cranes create change. If then, how light can our buildings be? Should we always look 30 years ahead as we build buildings? Other than the changes in the interior purposes or design, can the physical property of the building itself be changed as well? And will the city require this change?

We cannot say for sure, but many spaces nowadays need to be lightened and be capable of dealing with changes. In contrast to how supply has already exceeded the demand, and how the focus and movement of contents have become much more diverse and fast-paced, architecture still remains heavy as ever. Requests for remodeling and interior design have risen compared to the past, and many buildings that are still usable disappear because they cannot meet a certain floor area ratio or a specific purpose. I was having lunch recently at a restaurant in Seongsu-dong, and upon noticing its elegant interior, I probed deeper and found out that the restaurant kept the previous cafe interior intact. Perhaps it was to reduce dismantling costs and labor, but it was where I experienced a sense of spatial lightness. Just like Cedric Price's *Fun Palace*, we wish to create an architecture that is light and malleable for everyday use.

city built by countless number of people touches us deeply. The life and philosophies of various regional residents, and the sedimented urban architectures over many repetitions of the everyday create a massive spatial experience by itself. The climate and geography of the town and the space created from interpersonal relationships throw innumerable questions at our fixed conceptions of our everyday experience.

The trip to Cuba was especially memorable. The country may be poor in terms of numbers and figures, but during our time there at Havana, we were bombarded with the question, "So, are you truly happy?" Due to the lack of hotels, regular households function as accommodation facilities instead, and while not being exactly luxurious on the outside or the inside, the buildings displayed powerful color palettes and composition. We may possess cellphones and televisions that are superior to theirs, but they possessed leisure and an optimistic view on life. This lifestyle was seamlessly infused in every corner and space. Perhaps they may envy our lifestyles if we came to consider many other aspects, but this method of urban experience was a good opportunity for me to question again my conceptions on space.

6

With a port at its center, the scenery of Busan is dynamic. Especially, the massive scale of the containers at the port, the cranes, and the ships almost demand a sense of awe towards the accomplishments of civilization. However, the

7

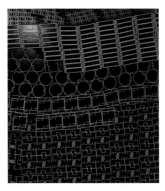

To execute these kinds of contemplations in the urban space, we have to be equipped with answers to these three practical questions: legal regulations, cost, and duration. Architecture requires a large amount of capital, and even if we do have the funds, most people in architecture grow 10 years older while creating a building. Things such as looking for a site, contacting an architect, and other matters related to various financial costs, tax handling, and management are all difficult even for architects like us. Furthermore, information is not readily available, and many people invest much legwork and wealth based on inaccurate judgments.

This is why apartments as a residential type — despite its negative properties such as the size of the apartment complex and its closed nature towards the city streets — is still an objectively attractive option in this Korean reality. As a standardized product that fulfills the basic residential function, an apartment is also attractive from a financial perspective because of its active market. I don't know what Le Corbusier himself would think of the current Korean scenery created by the apartments, but all would assume that his idea regarding the future of residence is most closely practiced — at least funtionally — in Korea. This is not a praise for apartments; rather, by analyzing the situation where we cannot but use apartments, we wish to bring about a change by applying the results from that analysis to future architecture. The smaller the site is, the more it tends to require manual work, and hence it is difficult to provide an estimate in terms of price and duration. However, to speak candidly, there are no such products in this world. It is a system that all participating bodies have to partake in the liabilities of unpredictability in an unreasonable way.

There is a need for a rational system that can provide quality space to more people via technological developments. Creativity and rational products for a good space are not antinomies; and as more people have access to more information, architectural culture will be able to eventually evolve.

8

As revealed from the beginning, we are challenging various aspects. We explore our diverse options multiple times a day, and wonder if we should just pick one to focus on for feasible reasons. On the other hand, however, we question ourselves again on whether the result will turn out well even if we choose to avoid those things that we were thinking on. We are still passionate and young, and we want to keep expanding our

thoughts further to create a better space.

One day, I saw myself working in a cafe, and I took a photo of myself. With technological advancements, my bag was supposed to get lighter, but being filled with things that I couldn't do without for work, the bag was heavier than before. Perhaps the era that we experience is in a process of a massive change, and maybe we are the inhabitants of this transitionary space.

When we were busy developing design automation, the match between AlphaGo and Sedol Lee was a hot issue. We were glad that our direction was aligned with the flow of this era, but we were also afraid of the image of the forthcoming world created by this massive change. Nonetheless, with an aim to build a more comfortable and better space within this changing times, we pursue an integration between creativity and technology.

9

What kind of an architect did we dream of, what kind of an architect are we, and what kind of an architect will we be.

As 'young' architects who just entered our 30s, it is still too early for us to provide a well-organized and clear account of the phenomena that have influenced us architecturally. We are filled with the passion to change the world through architecture, but we are still yet in the step of drawing the base sketches. Even at this point in time as we write this text, the countless phenomena and experiences that affect our architectural perspective are continually changing, such that the Boundless of 10 July 2018 who agreed to the editor's deadlines is different from the Boundless of 29 July 2018 who is now working hard to make the deadlines.

While these words that we have narrated thus far do represent a part of our aims, however, they were also expressed to state our purpose to the world in the form of an announcement that skips over all the dynamics of work and contemplation; and in that sense, they are not quite finished yet, but rather reveal our current status.

However, what is clear is that we are always open to change and that we are highly realistic. We also like observing and making discoveries, and we are favorably disposed towards the kind of complexity and uncertainties that arise from new relationships. Ultimately, we hope that these attempts will help to make our spaces to become a plane of sympathy and balance between the city and the human lives. In that sense, these images and narratives presented over 9 chapters are a combination of fragments that contain our experiences and ruminations that are yet to be organized. Because we desire our architecture to embody an open possibility so that it can continue to change and evolve in response to the complex and subtle relationships between the world and the human lives, we adopt this precisely unfinished conclusion as a part of our representation as we continue our search for the link between these fragments.

PROJECT

Design Automation

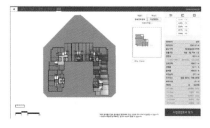

Technology and Design

The effort to automate architectural design through computers is not new. In 1971, George Stingy announced his shape grammar. This includes a logical structure that chooses and executes form vocabularies and the forms of its early state and its later state by rule of organization. As of now, organizations such as KPF, SHoP Architects, Aditazz, and Case Inc. which was recently acquired by WeWork are making attempts to connect computer science with architecture and the city. Beginning from IARC Architects' company research team Design Computation Group (DCG),

we have moved on to a research team Boundless-X and its spinoff startup Spacewalk to continue our research on architectural technology.

Range

Our primary aim is to provide project feasibility evaluations through design planning. Using urban data and AI-equipped architectural engine, we locate the land evaluation price, development method, financing method, and the post-development value. After this, the architect either adds or maximizes the value through creative solutions.

Orientation

We wish to create a technology that is meaningful to the masses. The existing professional system is effective for a large project scale. However, when the scale is small, there is a limit when it comes to using the original method. There are many such small-scale projects in the society. We wish to solve this through technology.

The small-to-medium scale residences are our targets. For apartments that are being marketed as standardized products, there is a significant collection of relevant data and indexes, and it is advantageous for the professionals to resolve this problem in an effective way due to its large individual sizes. While the small-scale residences occupy only 40% of the 1.6 trillion square meter residential area, these houses that may be small overwhelm in terms of their number as they represent 96% of the total number of blocks. However, since even the architecture law out of the 23 architecturally-relevant laws can change multiple times a year, most are excluded from information. One of the problems that we wanted to tackle overseas was the issue of land planning for farmers in Indonesia. 70 million Indonesian farmers live each day with less than 2 dollars. Because the amount that these people can afford is very limited, looking for a professional solution is almost impossible. We wanted to resolve these isolated problems through technology.

Architecture for Everyone

Many of our clients get worried when we propose the minimum design cost. A tailored luxurious suit may have its special value, but there should also be a generic ready-to-wear suit for everyone as well. A ready-to-wear suit can be created when a product that can be produced variedly in small amounts. We adopt this direction to build houses for numerous people isolated from the professional market by using technology.

We of Today

When Seoul Housing and Communities Corporation and Gyeonggi Urban Innovation Corporation were conducting their first project orientation with the local residents regarding the Street Housing-led Housing Renewal project as one axis of the urban revitalization project, we used a software because it is impossible to handle everything with professional consulting. Also, with KOICA's help, we were able to cooperate with a Vietnamese public organization to provide the prototype software for design automation that can effectively review the supply and investment of social residences that are accessible to everyone. After witnessing the recent collapse of a building in Yongsan, we developed a detector that detects deteriorating buildings that might be affected from neighboring constructions and supplied it to public organizations. It was an experience of contributing to the society through data. By matching our values with technology, we are striving towards things that are for everyone.

Greenlamp Library

A Space for Studying
Greenlamp Library is a premium reading room for students. Boundless participated in the design of this Greenlamp Library project from the planning stages and conducted the interior design and the branding process by cooperating with visual designers. By analyzing the studying patterns of the students, we designed six types of chairs and improved the space quality by installing a wide and comfortable lounge.

Participating from the Project Planning Stages
Greenlamp Library, which claims to be a premium reading room, is a project in which we actively participated in its spatial composition from the planning stages. First, we established its general brand image and planned its interior on this basis. Focusing on the intimate connection between brand and space, we established an integrative identity.

A Reading Room that Resembles an Open Library
While Greenlamp Library is mainly a study area for students, it was hoped that it would be appreciated more as a library than a reading room. The motif of classic library color codes, furniture, and lightings were reinterpreted in a modern way. Amongst them, the green lamps that are commonly found in old libraries were used as the foundation for the branding.

Six Types of Seats According to Study Pattern
In Greenlamp Library, students can choose between six studying options according to that day's study pattern as they move about freely between seats. Instead of a simple change in the form of the furniture, the seat types were divided by a spatial hierarchy and an opening and closing relationship. The original reading room followed the principle of one user per one seat, and thus it could receive only as many users as there are chairs. Through the diversification of seats, expansion of choice in space, and installation of a free-seating system, Greenlamp also has the advantage in terms of profit as the number of its users can now reach up to 130% of the total number of seats.

Comfortable and Wide Lounge Space
The fact that the number of users can now reach up to 130% of the total number of seats through the free-seating system is significant not only because of its increased profitability. As the profit could stay unchanged even if the number of seats within the same area size were to be reduced, it was now possible to provide a comfortable and wide lounge space to the users by expanding the shared space.

IT-Connected Feedback
We wanted to create a studious atmosphere not only by providing the physical hardware but also the software. As one example, under the slogan 'the place where time turns into talent', the users get reported their study hours from the digital board in the lounge and other information including using patterns in the form of a hardcopy monthly report. Like a ranking board in a game, the digital board displays information such as the

name of the student who studied the longest yesterday and the average number of hours that a successful applicant had studied, and a healthy, mutually-stimulating, studious environment that encourages effective study hours was created. Also, by using information of the preferred seats, the designer could make improvements for the consecutive branch libraries, and in this way Greenlamp is continuing to change.

Architect: Boundless Architects
Location: Repulbic of Korea
Program: library
Building scope: more 300m²
Interior finishing: gypsum board with paint, wood
Construction: Design Howl
Design period: 4 weeks
Construction period: 5 weeks
Photograph: Kyungsub Shin

Huam-dong Multiplex

Finding the Land

The Huam-dong Multiplex is a residence for an aged couple who are preparing for a retired life. The building owner sought for a small town that is a little more vibrant than a quiet outskirt and surrounded by a decent natural environment. Also, they wanted to live in a place where they could make enough profits to finance their retired lifestyle by means of a rental business. We began this project by searching for a land that meets all these conditions.
The Huam-dong site had met all these requirements, and due to its irregular shaped site, it was out in the market at a competitive price. Also, as an elevator that is connected to Sowol-gil was installed at the proximate street at the back, the area had the potential to become a commercial center with its increase of floating population. The urban scenery and city lights could also be enjoyed from the site's location, and we came to a quick decision to buy the land.

Entrance from the Proximate Street

Because the site had an incline, the main street levelled with the second floor. We planned for commercial facilities up to the second floor and created an inviting space there by clearing the center. The other task was to deal with the image of the building from the main street. Due to the setback regulation, a floor level difference was created, which caused the building's front to appear as its rear. By utilizing the floor level through a design language, we secured a sense of uniformity throughout the entire form and designed the rear to resemble the front.

Program Distribution

Without including the carpark, the first-floor area size is very small. Hence, we positioned a small kitchen and a space to welcome visitors on the first floor, while connecting the first floor with the basement as one space. The center of the second floor was emptied, and stores were positioned on each side. Three residential units of 36.3m² each were planned for the third floor, and a residential suite unit was designed respectively for the fourth and the fifth floor.

After Construction

After its construction, the environment began to change as the residents and store renters moved in. Visitors were putting up photographs of memorable scenes containing the building as the background on their social media platforms. People were sitting and chatting by a table on the shared space of the second floor that was open towards the street. People passing by were looking curiously at the building and its open shared space.

Architect: Boundless Architects
Design team: Kang Gayoon
Location: 43-5, Huam-dong, Yongsan-gu, Seoul, Korea
Program: housing and commercial
Site area: 216.78m²
Building area: 129.25m²
Gross floor area: 494.36m²
Building scope: B1, 5F
Height: 18.96m
Parking capacity: 5
Building coverage: 59.62%
Floor area ratio: 199.48%
Structure: reinforced concrete
Exterior finishing: brick, paint
Interior finishing: gypsum board with paint, wood
Structural engineer: Teo Structure
Mechanical engineer: Time tech.
Electrical engineer: Keukdong powertech
Construction: Taein Construction
Design period: 2015. 10 – 2016. 7
Construction period: 2016. 8 – 2017. 8
Photograph: Kyungsub Shin

Corner House

Balancing the Various Demands within a Small Site

Due to its commercial facilities in the lower half, the first floor of this building became crowded with people. A separate entrance flow line was needed for not just the first floor but also the second floor with the commercial stores as well as the basement, and a flow line and atmosphere that was separated from the commercial area was also demanded to protect the privacy of the residences of the upper half. Moreover, a parking lot had to be added according to legal regulations – it was not easy to respond to all these demands. To meet all these conditions within this small site of 132m² while not compromising in terms of design quality, we invested much energy in intense brainstorming. The solution began by utilizing the vertical level difference of the site.

Fulfilling Demands by Using the Level Difference

Because there were too many demands within a limited space, it was not possible to resolve the problem by floor area alone. The site's level difference had to be utilized, and we divided the flow line for facilities that required separate entrances by creating a semi-floor level difference. We placed the entrance for residences at the building's rear to protect privacy and secure the front image as a commercial facility. We created a large height difference in the basement entrance by using the incline of the corner, and this made the basement space more comfortable.

A Design That Considers Profitability

The rent price can change according to the program and the level despite the same area size. The rent of the first floor is the most expensive, and it was difficult to secure a large commercial area there due to parking space. By securing a large height difference in the basement entrance by utilizing the incline of Corner House, the rent was increased by 20%, and a larger profitability was achieved by connecting the first two floors via a skip floor.

Securing Ground Floor Parking Space via Tensile Pillars

After resolving the four entrances as requested, the parking space became too small. As the space between the ground floor entrance to neighborhood facilities and the basement entrance was made into space for parking, there were no longer space for structural pillars. Parking space was secured by using tensile materials instead of leading the pillars down to the ground. The force was transferred by the tensile material to other pillars.

A Facade that Considers Various Orientations and Levels

Being located on the corner of a five-way intersection, Corner House is visible from various angles and levels. Because the entire building to its pentagonal roof that made use of the legal restrictive lines is visible especially at the incline in front of the building, we used the dark but reflective masonry as the finish for the outer wall and the roof. The shiny property of the stone changes its color according to the direction of light. On the other hand, the commercial facilities of the lower half were finished with transparent glass in contrast to the upper half.

Bright and Entertaining Residential Space

Unlike the commercial facility of the open lower half, privacy had to be secured for the residential upper half. Also, it needed to receive enough sunlight for habitability, and along with such basic conditions, we planned to create an entertaining house that integrates spaces of various functions. While it looks closed from the outside, we created a pleasant entrance space by securing a bright shared space using sky windows and south windows, and thereby completed a house that embraces various spaces despite its small size using vertical zoning.

Architect: Boundless Architects
Design team: Sim Guedae, Kang Kyungguk
Location: 188-23, Nonhyun-dong, Gangnam-gu, Seoul, Korea
Program: housing and commercial
Site area: 132.3m²
Building area: 79.37m²
Gross floor area: 348.12m²
Building scope: 5F, B1
Height: 19.8m
Parking capacity: 3
Building coverage: 59.99%
Floor area ratio: 199.86%

Structure: reinforced concrete
Exterior finishing: stone, paint
Interior finishing: gypsum board with paint, wood
Structural engineer: Teo Structure
Mechanical engineer: Jung-in Eng.
Electrical engineer: Daekyung ENC
Construction: Taein Constructuion
Design period: 2013. 11 – 2014. 4
Construction period: 2014. 5 – 2015. 1
Photograph: Kyungsub Shin

CRITIQUE

To Boundless Architects

by Minah Lee

Note from the Author
In a way, I'm writing a critique on three individuals' works. I do not think that a separate critique for each submitted work to the Korean Young Architect Award is necessary. I brought up my memories of these three individuals with whom I had my first acquaintance 15 years ago, and with a refocused viewpoint, I read out the new signals that they've brought to my attention. We had an extended talk for a long time since. I thought I would be hearing the gloating of a prize winner, but instead I sensed both a feeling of instability of a prematurely-birthed infant and a state of exhaustion of an overly mature child at the same time. However, within them, there was also a contemplation of a profound depth as well as an exercise of rigor towards the essence of their work through acute questions such as, 'so where and how will we head on from here?' I had wanted to ask them if winning an award really matters, but I couldn't. In fact, when I was acting as the examiner for the Korean Young Architect Award, I was hoping deep inside that these three would soon appear. I wrote this text directly and solely for Boundless. This text also made me reflect a lot upon my architectural work.

Year 1
When I first received news that these three were working together, my memories of Jihwan Lim wasn't as clear as my memories of Jooho Moon and Sunghyeon Cho. Even before the official announcement, some of the younger people in the office who were following their works through SNS had been telling me how big of a news it is for them to come together, but I wasn't really deeply affected. I came to meet Im in the first semester of 2003, and I met Moon and Cho later in the second semester. The studio for the first year's first semester was automatically assigned by student number, and Lim became my student. The first-year architectural students usually come into the design class wearing a provocative expression towards their lecturers, being filled with an adorable sense of pride for their accomplishments in becoming students of Seoul National University. The teaching method may be quite different from their high school years, but being half curious about what architecture is, and being half filled with self-confidence, students come to engage in a light psychological warfare with the teachers. Just like Moon and Cho in the second semester, I remember Lim was someone like that.
In the second semester, students could assign themselves to a studio of choice. Cho and Moon, whom I've noted down as unique individuals in the neighboring class during the first semester, Choined my class. The first-year's second semester in the architecture department is somewhat of a crisis period. It is right after the first semester where students come to experience a significant deprivation of sleep and the burdening stress of creation, and also the holiday break which they couldn't properly enjoy partly due to that semester's aftereffect. The students in the second semester no longer address their teachers with an innocent and adorable expression but harbor an expression where it shows that they've been working hard convincing themselves of their major choice. Amongst them, however, Moon's focus was extraordinary. The unique sketches that he was displaying seemed to be a selection out of an even greater number of sketches, and when his thought delivery was not sufficiently clear, he insisted on going back to the beginning, thus stealing away other students' time for questions. I have a strong memory of our conversations that went beyond the tutorial materials. Cho was like a neighborhood friend who came back after studying at Pohang University of Science and Technology. He wasn't the punctual kind in terms of handing in assignments, but his enthusiasm and his confident and somewhat irrelevant questions – questions that made me wonder about the limits of his interest – were provocative and interesting. Other students who were listening into our conversations comforted themselves from being burdened with design assignments while finding Cho's artful and flexible brain fascinating. In contrast to them, Lim was sharp, but also quiet. I think he intentionally made his acute questions sound crude. He was different from other students who were always agreeing or flattering their lecturers. With an astute ability to control one's own speed and energy, he did not want to go along with the flow of the class. I remember that these three individuals shared a kind of an unyielding nature.

Then a few years ago, some acquaintances who were also fans of their work came to me with a famous magazine to show me how well Boundless was doing, and I came across a photo of Moon sporting a goatee and Cho with a robust and healthy-looking face. Lim's unchanged, humble smile made me wonder about his role among these three, and I became seriously concerned regarding the boss-like appearances of Moon and Cho.

Office

I had some preconceptions when I first heard that these three had formed a team. I had doubts if the trio could truly effectively draw upon each another's strengths. I also had questions about their team name choice. Is it about the essence of architecture, in the general sense of boundary, or is it a marketing strategy to promote their willingness to do any kind of work? Or is it a statement that boundary should disappear? Or is it simply a choice for a memorable name? Regardless, their concept for 'something without something' is quite old and cliched to be picked by those three. Also, I was somewhat disappointed in their unabashed desire to market themselves, being a team formed by graduates from Seoul National University. However, I was also curious about this group who was already announcing its construction project without taking any delays from its stepping into reality.

The bigger purpose of meeting them was actually to observe their office. The office located in the corner of a building at Seongsu-dong was rather unexpected. As if they were just temporarily renting a seat of a moving train car, to them, the physical space was not as important. Lacking any obsession towards a personal physical space, Moon leisurely guided me to a small conference room as though I was being led to a dining car of a train. As someone who was tied to the idea of a personal space in the office and used to being surrounded by the weight of various things, the restrictive collection of records, and the landscape of piled up books that might collapse the floor, I was intimidated by the boundless nature of this office. I felt like an uncivilized individual within this un-locational unfamiliarity. The meaningless boundary that these individuals were driving at was redefining the conventional space and instigating a synergy with the neighboring professionals of other fields. This helped me to understand the present identity of Boundless. In summer of 2016 when it turned three years old since its founding in 2013, Boundless faced a self-created crisis. Its sales, sites, and the number of staffs were rising, but the three individuals asked themselves if they were by themselves happy. Then they asked if they helped to make the world a better place. They must have

thought of these questions as they first began this team. As they were forming a team in 2009, they assured themselves that there are numerous other possible paths even if the team doesn't work out. They continued their reflections on 'so what are we dedicating ourselves towards' throughout their growth, and they rigorously challenged each other on how these contemplations can be translated into practical values that are beneficial to the society, building owner, and the architect. Simultaneously with their effort to refurbish their strategy, the three sought to mature as individual architects. A year later, I heard from my acquaintances who were following their progress that the Boundless had dissolved. Boundless was a model and a most popular workplace for the younger generations. While this news brought a sense of loss for the symbolic worth that Boundless had represented, it also made me think that the decision would have been possible if they truly sought to transcend from all boundaries. The three must have put themselves through a short period of despair and self-evaluation amid their busy schedules. With the passion and the skills that they have thoroughly exposed to one other over this period of transcending and self-scrutinizing, the three individuals gathered again with a rebolstered foundation of their own.

Text, Speech

The three have sent numerous texts of various types and lengths. I analyzed the words that these three often use in their sentences and focused on the concepts that the three identified closely with in their words and speeches. There is an unpolished monotone in their language, but I was also able to read their accurate perception and discerning insights towards social change. Their writings also find commonness in their adoption of a dry style where they refuse any kind of linguistic amusement.

Moon's essay is filled with concerns that he might lose track of the essence of space which he ultimately desires to investigate during the design process. Like the metal bits that gets attracted towards a magnet, even after mentioning things regarding development profitability and projected revenues, he immediately returns to emphasize the function of the architect to pursue good space. While the three do not copy after the sensitive metaphorical prose of the previous generations of architects and stick merely with words that are related to the tasks at hand, however, their unpolished writings seem to embody the future. Meanwhile, Moon wrote, 'because we are architects, we like to empty things'. Ultimately, these kinds of reflective writings and the vocabulary leap that architects choose and write reveal where they are in their contemplation. While architectural writings can be quite hypocritical, it is also a confession that can make one self-reflect time to time. Disregarding the fanciful vocabularies, the texts of Boundless simply repeat and reemphasize the words that come forth presently as most urgent to them. This is different from the empty formalities of the previous generations of architects who wished to bring about an intellectual reaction by using their created conceptual words as a source of creative inspiration. As someone who is used to architectural writings that are based on diverse experiences and inspirations from reading short stories, I find the mundane prose of these three individuals that express merely the bones with no flesh quite disturbing.

In Cho's speech, there is a kind of a mechanical tone to it, as though it was developed to explain what he does in a well-organized way. His eloquence makes one cheekily wonder if he programed the meanings and the tone for the most effective delivery. When Cho was explaining the village masterplan automation project to resolve the land planning issue of a farming village in Indonesia, however, his voice became relatively thinner and unstable. What is it that makes Cho tremble? It must have been difficult for him to candidly express to me his desires to put himself into a task with a definite public value, but I could see that a sincere interexchange between the languages of persuasion and confession were taking place without him being aware of it.

After getting to read Lim's lately submitted text, I was finally able to understand my problem. What did I believe in architecture when I was in my 30s? How did I live the time that Lim is living today?

Work, Beyond Work
My first acquaintance with the construction project was from a peer review at the monthly magazine *SPACE* a few years ago. Despite it being a blind evaluation, I was immediately aware that the work was done by the three individuals, and I was even reminded of the three managers of the architecture offices that these three had worked at respectively. I did not select their work. Despite that their three design intentions towards development profitability, reasonable spatial utility, and formativity was well visible, because of my assumption that the work came from them, I made a subjective choice that goes in opposition to the nature of a blind evaluation. This was because I did not see that they went all the way. I had hoped that they would engage in an intense debate or a challenge against each other's sensitivities in regard to the essence of space during the planning stages. I was beset with a complex mix of suspicions on why they had stopped midway, wondering if it was out of a compromise among

the three, or because they were not aware of the detail. I only got to know through an interview that they in fact go through a series of debates and mutual persuasions every time. The three had created a rule which states that the idea must be reviewed if at least two among the group express opposition; however, if one among the group has a very strong opinion for the idea, the rule demands that the other two must review their opinions instead. But I wonder: the three may control each other in this way, but can architectural space always be filtered via a safety rule of choice? Perhaps, there are more cases where one finds oneself closest to the foundations of architecture as one bears the heavy burdens of decision-making by oneself. In a way, these three individuals were unjustly exempted from the review for being my students, and my decision went askew from the intentions of the peer review system of *SPACE*. However, I hoped that the three would not think it so important for an architect to have their work put up on a magazine.

The so-called intellectualism of this trio which has accomplished works that range from its early urban-lifestyle residence project to its current botanical garden project is clearly a very important factor in validifying the team's credence to other external observers. However, as one moves onto a kind of a self-restriction where one binds oneself under the thought that an elite office of Seoul National University graduates must always be able to create something distinguished, one becomes increasingly reliant on exaggerative words and texts in describing every project. I did not want to see a repetition of words in their construction projects, and I also did not expect them to show a clear distinctness that necessitate explanation. The architecture of the three individuals possesses a calm first appearance and a unique texture. It also possesses a meticulous proportion, clever designs, and warm humor. It also has objects and landscape that make me envious. Why did we not have this kind of conversation before? The Corner House in Nonhyeon-dong and Huam-dong Multiplex may have now become the trio's representative works, but I wonder if the values that are emphasized here – the value of land, utility of space, and securement of the public – are also not things that are being intensely pursued by other architects as well. The core expressions that keep appearing in the three's texts – 'creativity', 'maximization', and 'attributing value' – surely do not belong uniquely to Boundless alone as their intrinsic design principles. Even so, what is that target of dedication that makes these three individuals to strive themselves to the end in terms of these basic elements?

With unique ideas and intellectual sensitivity, the three engaged with the most practical, public spatial by-product – a reading room – in a creative way. By drawing out an interesting anthropological graph and graphics, they utilize the desires of parents and students alike, motivating them in a novel way and answering to them with a most effective space. It is difficult to divide where the project fits more of a product and where it fits more of a social contribution, but because of this novel, awe-inspiring evolution of a reading room, I developed a desire to see public interest variants of this Greenlamp Library project under the hopeful expectation towards the countless possibilities of such change in the chaotic city. If Boundless and its partner team Atostudy were to step up to propose a 'clean' variant project of the Greenlamp Library by systematizing the space of garbage collection center and the price of used tissue paper, providing glow-in-the-dark shirts and warm winter uniforms of good designs, respectfully positioning low-income elderlies at the front of the resource recycling effort, and designing the trolley sides of these elderlies with a graphic contribution that is more eye-catching and beautiful than the ads posted on city buses, I think that the aim to 'protect human decency' that the trio had ambitiously proposed might become amicably fulfilled.

By dealing with realms that may or may not be automated through design automation, the computer expert Cho asserts that this skill is only applied to regulation reviews that involve less human judgment. Also, he provides this information and software to many people who are excluded from the professional market so that they too may have the chance to consider building a house. I do not find the automation so interesting, however; instead, I find it more fascinating how Cho had managed to configure a mechanism that makes one keep wondering on what the essence of architecture is. He must have first asked Moon and Lim for their thoughts on design automation, and the three must have spent a day contemplating on questions like 'what is the role of an architect'.

The way to continue
The depression sensed in their boss-like countenance can no longer be found from them. Because they are so entrenched with things to investigate with their current approach, the three haven't had time to consider studying abroad, or even to consider making preparations to apply as a public architect. Instead, they have made families, and they must have had encountered at least once or twice the world's vanities and the deep materialistic desires in themselves. They must have experienced the pretensions of a professional, the troubles of a businessman, and

a sense of despair and defeat. Also, they must have witnessed how the smallest architectural practice is meaningful in that it contributes to one's authentic growth as a part of world-living experience.

Indeed, they must have realized that even though boundaries may not exist, a certain critical point that can put us in despair definitely exists. Exceptions always exist in issues regarding space and location, and they must have seen how wide of an undesignable grey area that resides within architecture and the urban city is.

The architecture that we strive for is ultimately material, and because it holds both physicality and textuality, they might have told themselves once that it is fine to make decisions based on simple intuition and disposition. Due to a strangely innate belief towards the purity of architecture, they might be reminded of how they had to struggle inside while confronting with the ethics of architecture. They might also have had the desire to build a private individualized order that does not belong to any classification. There must also have been times when they wanted to leave the group, considering how times of individuals are not equivalently interchangeable or shareable.

Perhaps, they might have become more sensitive towards the world's opinion after being awarded, such that they might now wish to revamp their images in expectations of social media coverage and interviews. Perhaps, they might have wanted to sincerely express their dreams to be an entity that contributes to society while acting as a developer. However, they must also be aware of the fact that service loses its meaning once it is linguistically expressed in architecture.

To Moon who graduated the college with honors, to Lim who was awarded for his graduate project, and Cho who has almost finished reading all the books on Donald Trump? Do you all still remember the passion of your student years? Where are they three at right now? (Translated by Keunho Hong)

Minah Lee majored in Architecture at Seoul National University, where she also received a Master's degree in 1991. Lee also received Master's degree from the Berlage Institute in Amsterdam. After working at Space, H.Min architects and associates, and Kiohun, she now runs her own office, the laboratory of architecture Hyupdongone.

남정민

서울과학기술대학교
OA-Lab 건축연구소

Jungmin Nam

Seoul National University of Science and Technology
OA-Lab (Operative Architecture Laboratory)

남정민
연세대학교 건축공학과를 졸업하고
하버드대학교 GSD(Graduate School
of Design)에서 건축설계 석사
학위를 받았다. KVA(Kennedy &
Violich Architecture), OMA(Office
for Metropolitan Architecture),
Safdie Architects 등 다양한
사무소에서 인턴과 실무 경험을
하였다. 현재는 서울과학기술대학교
교수로 재직하며 OA-Lab
건축연구소를 설립하여 활동하고
있다. OA-Lab을 통해 실험적인
건축을 추구하고 연구와 실무를
오가며 작은 제품에서부터 건축물에
이르기까지 다양한 범위의 작업을
하고 있다. 관찰과 실험에 기반한
디자인이 우리의 일상과 사회 속에서
삶의 경험을 담고 긍정적으로
작동하여 구현되는 것을 목표로
하고 있다. 하버드대학교에서
졸업논문상 파이널리스트 및
추천장을 받으며 졸업했고, 이후에
AIA보스턴건축가협회의 주택공모전
대상, AIA국제지역건축가협회 대상
등을 수상하였다.

Jungmin Nam
is an architect and educator. He is
currently teaching as a professor at
Seoul National University of Science
and Technology and conducting
design research as a founding
principal of OA-Lab (Operative
Architecture Laboratory). Previous to
his own practice, he has gained his
architecture education at Harvard
University, Graduate School of
Design, graduating with the Letter
of Commendation and experienced
his professional practice at KVA
(Kennedy & Violich Architecture),
OMA (Office for Metropolitan
Architecture) and Safdie Architects.
By focusing on experimental design
ranging from small scale products
to large scale architecture and by
believing design should engage with
our society and culture, Jungmin
Nam is trying to bridge between
academic design research and real
world practice, through which he
believes that he can contribute
architecture culture and society.
He received numerous awards,
including Harvard GSD's Thesis
Award Finalist, AIA International
Regions' Honor Award for
Architecture and BSA/AIA's Housing
Competition 1st Prize.

에세이

사유와 공공의 경계에서
발생하는 가치

남정민

남정민

I

도시를 걷다 보면 벽, 보도블록 등의 틈새에서 피어나는 식물을 종종 볼 수 있다. 별 관심 없이 지나치던 이런 풀들을 주목하게 된 것은 도시의 잡초들이 특정 국가나 지역에 국한되지 않는 현대 도시의 보편적 현상이라는 것을 느끼면서부터다. 서울뿐 아니라 유학 시절 머물던 케임브리지와 여타 다양한 도시에서도 길을 걸으면 종종 구석구석에 있는 틈바구니에서 이런 잡초들을 마주할 수 있었다.

바닥으로 낮춘 눈높이에서 우연히 바라본 잡초들은 더 이상 미미하게 흩어진 식물이 아닌 그 자체로 하나의 생태계를 구성하고 있는 듯 강한 존재감을 보였다. 이런 식생들이 기존의 고정관념 이상으로 역할이 크지 않을까라는 생각은 이후 식물학자 피터 델 트레디치(Peter Del Tredici)가 하버드대학교 GSD에서 학생들과 함께 조사한 자료를 통해 간접적으로 확인할 수 있었다. 그의 조사에 따르면 매사추세츠주 섬머빌(Somerville)의 경우 도시 표면의 약 9.5%를 잡초들이 점유하고 있고, 이 면적은 사람들이 관리하는 공원의 면적을 넘어선다.

도시의 잡초들은 그 면적만큼이나 도시 환경에 기여하고 있다. 이들은 포장도로에서 도시의 열을 감소시키고, 비가 올 때 흙을 붙잡으며 물을 머금고, 오염물질을 제거하며, 작은 생명체의 생태계도 된다. 자연발생적으로 도시에 자라나는 이런 잡초들은 그 나름의 생태계를 구성하는 패턴이 있다. 나의 건축 과정에서 도시의 잡초는 제거 대상이 아닌 미리부터 고려하여 포용해야 할 건축의 중요한 일부이다.

남정민

2

흔히 서울을 아파트공화국이라 부르지만, 1970-1980년대로 거슬러 올라가면 서울 시민의 주된 주거 유형은 단독주택이었다. 아파트 단지가 되지 않고 남아 있던 거주용 도시 조직들은 단독주택에서 다가구, 연립-다세대 주거로 급속히 유형을 바꿔왔고, 오늘날에도 여전히 단독 및 다가구, 다세대 주택은 서울시 주거 유형의 50% 이상을 차지하고 있다. 즉 서울 시민의 50% 이상은 최소한의 법정 조경 면적도 적용되지 않은 주거 밀집 지역에 거주하며 일상을 보낸다고 할 수 있다.

공원 및 공지뿐 아니라 최소한의 조경 공간도 없는 물리적 환경에서 개인의 경제적 욕망에 따라 대지경계선까지 건물로 가득 채운 곳에서도 사람들은 틈새 공간을 찾아 자연이 들어설 자리를 만든다. 자신의 담장과 대지 경계 안의 틈새 공간을 찾아 화분과 화단을 만들어 식물을 키우고, 담벼락을 따라 나 있는 길의 형태와 폭, 이를 사용하는 사람들의 동선 사이에서 틈새 영역을 찾아 화분을 놓고 텃밭과 식물을 가꾼다.

건축가의 개입 없이 거주자의 욕망으로 일상에서 행해지는 자발적인 조경은 물리적 콘텍스트와 개인 및 공공 사이의 보이지 않는 역학관계 속에서 자연발생적인 식생처럼 자신만의 패턴을 가지며 형성된다. 이를 통해 무표정하던 2차원적인 담벼락과 건물의 입면에는 입체적인 식생의 깊이가 더해지며 길과 건물 간의 대화가 시작된다. 담벼락을 따라 지극히 사적인 목적으로 형성되었던 식생은 지역 사람들이 골목을 걸으며 공유하는 풍경이 되고 작은 공원이 된다. 일상의 건축에서 표면은 건축물과 도시, 개인과 공공의 치열한 접경지에 놓이고, 표면이 입체적 깊이를 가질 때 여기에 형성되는 틈새 공간은 도시 안에서 새로운 가능성을 품을 수 있다.

남정민

3

건축은 역사적으로 자연을 극복하고 대립하며 발전해왔다. 건물이 지어지고 도시가 형성되면서 자연이 있던 곳에 건축물이 들어섰다. 하지만 건축물과 자연의 대결 가운데서도 사람들은 일상에서 자연을 곁에 둘 방법을 다양하게 찾아왔다. 도시화를 피해 남겨진 산과 같은 자연을 찾아가고, 도시 안에 공원을 만들어 자연을 일상에 들여놓고자 했다. 개인의 삶에서도 정원과 화단을 만들거나 그조차 용이하지 않을 경우 화분을 통해 식물을 키우고 가꾸며 일상에 자연의 자리를 만들어둔다.

하지만 자연을 곁에 두려는 일련의 행위들은 주로 건축의 영역 밖에서 방법을 찾아왔다. 건축이 자연을 밀어내고 그 자리에 인간을 위한 공간과 장소를 만드는 데 집중해왔기 때문이었을까? 건축물을 계획할 때 자연은 대부분 법정 조경 면적 안에 갇혀 건축물과 구분된 대지의 빈자리에 적용되거나, 건축 이후에 덧대어지는 부수적인 요소로 다뤄져왔다. 건축이 행해지는 곳에서 자연은 건축의 필수 요소라기보다 건축 외적인 추가 요소로 소극적으로 다뤄졌다.

하지만 세계 곳곳에는 적극적으로 자연과 건축의 통합을 시도한 전통적인 사례들이 있다. 대표적으로 북유럽지역 건축물의 잔디지붕은 자연과 건축의 관계를 보다 통합적으로 바라볼 수 있게 한다. 북유럽의 잔디지붕은 그 지역의 기후 특성, 목재에 기반한 구축방식과 함께 지붕을 구성하는 필수 요소로 19세기 무렵까지 널리 사용된 지역 건축 방식 중 하나이다. 자연을 건축 구성의 주요 요소로 고려해 건축물의 외부적 형태와 기능 및 구축에 통합적으로 적용한 사례이다.

4

18세기 무렵 영국의 콘월(Cornwall)지역의 채광 산업을 이끌며 산업혁명에 기여했던 엔진하우스(Mining Engine House)는 19세기를 지나며 지역의 채광 산업 몰락과 함께 건물의 껍데기만 남겨진 채 덩그러니 버려져 있다. 오늘날 유네스코 세계문화유산으로 지정된 이 엔진하우스들은 그 규모와 개수가 한때 번성했던 채광 산업만큼이나 방대해 지금도 콘월지역 여기저기 대지에 불쑥 솟아 있는 것을 볼 수 있다.

콘월 엔진하우스는 애초 사회·문화적 목적 혹은 자연에 대한 존중과는 전혀 상관없는 건축물이다. 오히려 자연을 소비 고갈하고 산업 발전이라는 목적과 기능에만 충실했던 건물로, 채광에 적합한 장소에, 채광에 필요한 엔진의 크기와 규모에 따라 기능만을 고려해 세워진 건축물이다. 하지만 100여 년이 지난 지금, 그들은 마치 그 지형이 생길 때 땅이 빚어낸 랜드마크처럼 자연과의 대비 속에서 그 일부로 자연스럽게 서 있다.

콘월 엔진하우스는 이 지역 돌을 사용하고 엔진하우스 본연의 기능에 충실하도록 필수 기능만 최소한으로 담은 건축물이다. 때문에 굴뚝과 단일 매스 같은 필요 최소 단위만 활용한 간결하고 이성적인 디자인 형태를 가진다. 이로써 자연과 인공 축조물 간의 어울리는 대비를 보여준다. 이는 자연만 홀로 있을 때 성취하기 힘든 특별한 분위기다. 도시는 필연적으로 건축물과 자연을 함께 품을 수밖에 없다. 콘월의 엔진하우스는 건축과 자연이 함께했을 때 만들어낼 수 있는 새로운 관계와 가능성을 다시 한 번 생각하게 해준다.

남정민

5

1990년대 후지모리 데루노부(藤森照信)가 설계한 도쿄 교외의 민들레의 집(タンポポ・ハウス) 하우스를 비롯해 2000년대 스테파노 보에리(Stefano Boeri)가 설계한 최초의 수직숲 아파트 빌딩 보스코 베르티칼레(Bosco Verticale)에 이르기까지 자연과 건축을 통합하려는 시도는 현대에도 지속적으로 명맥을 이어가고 있다. 북유럽의 전통 건물에서 건축과 자연의 통합을 주로 지붕을 두고 시도했다면, 20세기 후반 이후 근래의 시도들은 그 통합을 건축물의 입면인 수직의 외벽까지 확장하고 있다.

근래에 수직녹화를 하나의 사업 아이템으로 적용하여 모듈화된 화분과 관수시스템을 통합한 다양한 제품이 이미 시중에 많이 나와 있다. 국내에서도 이 제품군들이 다양하게 관공서 건물 및 상가 인테리어 등의 수직녹화에 활용되고 있다. 하지만 이런 녹화 시스템 대부분은 식물과 건축 간의 통합적 관계에 대한 고민이 부족해, 2차원적인 수직 평면을 식물로 채우는 기능만 고려하며 건축과의 통합보다는 건축에 덧대어지는 별개의 독립된 제품으로 존재한다.

이에 반해 후지모리 데루노부가 보여주는 태도는 늘 대비되는 관계에 놓여왔던 건축과 자연을 보다 통합적으로 재정립하는 시도가 엿보인다. 그의 건축은 건축설계에서 조경으로 치부되며 소외되어왔던, 식물로 대변되는 자연의 존재를 건축의 일부로 인식하며 건축의 범위를 자연으로까지 적극 확장하고 있다. 아직은 현재 진행 중인 실험 과정이지만, 이를 통해 앞으로 건축이 가질 수 있는 가능성을 생각해보게 된다.

남정민

6

루이스 설리번이 설계한 개런티 빌딩(Guaranty Building)의 부분 확대사진을 처음 봤을 때의 당혹감은 풍부한 장식 때문이었다. 이 건물을 루이스 설리번의 '형태는 기능을 따른다(Form follows function)'라는 너무나도 유명한 개념에 기반한 대표적인 건축물 중 하나로 알고 있던 까닭에 건물 구석구석을 수놓은 아름답고 풍부한 장식은 예기치 못했던 반전이었다.

기능에 기반해 고층 건물을 수직적으로 4개의 영역으로 구분하고, 외관에서도 철골조의 구축 질서가 그대로 입면에 반영된 초기 고층 빌딩인 이 건물은 지극히 기능적인 원칙을 따르는 건축물의 구성을 보여준다. 하지만 이전의 고전 양식의 건축물 못지않게 풍부한 장식을 두르고 있다. 자세히 보면 이전 세대와는 다른 이 건물만의 고유한 새로운 장식적 패턴으로 표면에 깊이를 더한다.

'형태는 기능을 따른다'라고 하면 무미건조하고 획일화된 국제주의 양식의 건물로 오해할 수 있지만, 루이스 설리번이 제시한 기능은 사회·문화적 기능까지 포함한 훨씬 더 풍부한 것이었다. 얼핏 기능에 반하는 것으로 오해를 살 수 있는 건물을 뒤덮고 있는 테라코타 장식으로, 이 건축물은 이전의 전통 건축과는 다르지만 정겨운 표정을 가지며 도시의 풍경에 스며든다. 초고층이라는 낯설고 이웃을 압도하는 새로운 이방인이 도시에 들어섰지만, 친절한 표정을 지으며 길과 주변 이웃을 향해 인사를 건넨다. 그 인사 덕분에 이웃(건물)들은 스케일에 압도되던 두려움을 떨쳐내고 함께 도시의 구성원으로 받아들일 수 있었을 것이다.

The Seagram Building uses the steel structural grid of
the skyscraper to create a vertical affect, by attaching
a series of decorative I-beams to the envelope that
prioritize the vertical lines of the structure over the
horizontal floor plates.

The I-beam sections attached
to the façade are part of the
prefabricated window units
that make up the skin
— revealed by the horizontal
gap between I-beam sections
from floor to floor.

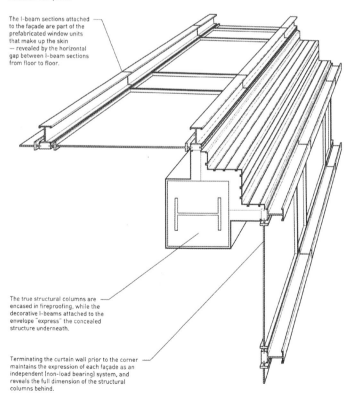

The true structural columns are
encased in fireproofing, while the
decorative I-beams attached to the
envelope "express" the concealed
structure underneath.

Terminating the curtain wall prior to the corner
maintains the expression of each façade as an
independent (non-load bearing) system, and
reveals the full dimension of the structural
columns behind.

7

차분하게 검은색을 띠고 서 있는 시그램 빌딩(Seagram Building)
의 수직 멀리언들은 얼핏 장식을 배제한 질서정연한 모습이다. 그
러나 디테일 도면에 드러나듯 이는 구조 역할보다는 사실 장식적
효과에 더 가깝다. 수직 멀리언 장식이 없는 시그램 빌딩을 과연 동
일한 시그램 빌딩이라 할 수 있을까? 단정하고 계획된 외관의 이
건물은 장식과 건물의 관계를 되돌아보게 해준다.

이후에 경험한 건축물 외피와 장식에 대한 다양한 해석은 장식 그
자체뿐 아니라, 건축물의 외부를 구성하고 건축물이 외기의 변화
에 대응하면서 구축 방식과도 밀접하게 연관될 수 있음을 보여준
다. 장식을 통해 전달되는 건물의 외피는 건축물의 안과 밖을 매개
하며, 직간접적으로 외부 공간에 대한 내부 공간의 경험을 제공하
고, 내외부 사이의 관계에 영향을 주고 있다.

파시드 무사비(Farshid Moussavi)의 『장식의 기능(The Function
of Ornament)』은 기능의 의미를 구조 혹은 프로그램과 같은 엔지
니어적 기능에 국한하지 않고 사회·문화적 기능과 건축물과 도시,
건축의 외부와 내부 간의 의사소통으로 의미를 확대해준다. 이는
건축물의 외피를 얇은 껍질처럼만 바라보던 제한된 시야 또한 함
께 확대할 수 있는 계기가 되었다. 표면의 깊이라는 말이 이 책에
직접 언급되지는 않지만, 이 책에 소개된 수많은 성공적인 건축물
은 결국 표면에 깊이를 가짐으로써 건축물로서의 또 다른 기능, 즉
건축물이 들어설 때 피할 수 없는 외부와 주변의 이웃들에 대한 역
할을 완수하고 있다.

LEARNING
FROM
LAS VEGAS
Revised Edition

Robert Venturi   **Denise Scott Brown**   **Steven Izenour**

8

『라스베이거스의 교훈(Learning from Las Vegas)』은 일상의 건축과 주변 환경을 바라보는 나의 태도와 관점에 많은 영향을 끼친 책이다. 이 책을 통해서 평범하게 지나칠 수 있는 일상이 새로운 발견과 배움, 관찰의 대상이 될 수 있다는 사실을 재인식했다. 이 책은 일상에서 얻은 배움이 때로는 얼마나 큰지 교훈을 주고 있다.

이 책에서 저자들은 라스베이거스를 대상으로 일상의 건축과 상업 건축에 대한 관찰과 분석을 통해, 이들 건축이 가진 역할과 가치를 재발견한다. 여기서 당시의 지배적이던 모더니즘에 대한 대안으로서의 가치를 제시한다. 저자들은 일상의 건축을 통해 기존에 확립된 건축계의 지배적인 가치를 재평가하고, 일상에서 찾아낸 건축의 가치를 다시 한 번 분석하고 정리해 새로운 대안으로 제시하는데, 이런 일련의 과정은 나에게 시사하는 바가 컸다.

이 책을 통해 한편으로는 스스로에게 질문하게 된다. 저자가 1960-1970년대 라스베이거스에서 건축의 가치를 재발견하고 대안을 제시했던 과정이 오늘날의 한국 건축계에서도 일어나고 있는가? 혹은 일어난 적이 있는가? 이 책의 저자들이 보여준 접근 방식과 태도는 시대와 장소를 초월한 보편성을 가지며 오늘의 한국에도 시사하는 바가 크다고 본다. 이 책에서 건축을 관찰하고 분석하는 과정과 태도는 지금의 나에게도 여전히 건축을 대하는 길잡이가 되고 있다.

남정민

9

우리가 직접 들어가서 경험해본 건축물은 과연 얼마나 될까? 도시
의 수많은 건축물 중에 앞으로 직접 사용하며 경험하게 될 건축물
은 과연 얼마나 될까? 우리가 도시, 도시의 장소들 그리고 그 안에
서 건축물들을 어떻게 경험하는가를 물리적 환경으로 좁혀서 생각
해보면, 그 경험들의 상당수는 우리를 둘러싼 건축물의 입면들로
구성되어 있다.

책을 읽고 그림을 보고 음악을 듣고 영화를 볼 때 사람들이 해당 매
체와 내용을 체험하는 과정은 비교적 유사한 방식으로 공유될 수
있다. 이에 반해 건축물의 경우, 그것을 실제로 사용하는 경험과 표
면만을 경험하는 것으로 나뉘게 된다. 건축물의 안팎을 아우르며
일상에서 겪는 다양한 경험은 실제로 그 건물을 사용하는 소수의
사람에게 국한된다. 반면에 오히려 건축물이 속한 도시나 이웃을
공유하는 다수의 사람은 그 건축물을 표면으로만 경험하게 된다.
이는 선택의 여지가 없는 일방적인 경험이다.

건축물은 개인의 욕망과 소유로 설계되고 만들어질 수 있지만, 결
국 표면을 통해서 다수인 공공의 삶에 영향을 미친다. 건축물이 외
부와 공공과 소통하기 위해서는 도시가 보다 쾌적하고 즐거운 물
리적 환경에 대한 경험을 제공해야 한다. 또한 건축물의 내부 공간
뿐 아니라 표면의 역할도 크다. 서울처럼 밀도가 높고 용적률로 가
득 찬 도시에서 표면이 할 수 있는 역할은 오히려 더 커질 수 있다.

건축이 가진 공공의 역할에서 표면은 필수적인 요소 중 하나이다.
사유와 공공의 경계에는 늘 건축물의 표면이 놓여 있다. 이에 대한
다양한 가능성을 탐구해볼 때 우리 도시는 보다 나은 환경을 가질
수 있을 것이다.

리빙 프로젝트
Living Project

일상의 건축화, 화분에서 스트리트 퍼니처까지

건축물의 표면은 건물과 도시 영역 사이에서 개인의 욕망과 도시의 공공성이 충돌하는 치열한 경계선으로 일상에 존재해왔다. 그리고 우리는 이 건물들의 표면이 만들어낸 도시 공간(길, 광장 등)을 통해 건축물을 경험한다. 서울 같은 고밀도 도시일수록 건축 표면이 가진 공공의 가치는 역설적으로 더욱 중요할 수밖에 없다.

건물은 개인의 욕망으로 지어지지만 표면을 통해서 도시와 대화한다. 그 표면이 2차원의 평면을 넘어서 깊이를 가질 때 더 많은 이야기를 함께 나눌 수 있다. 서울 근린생활 지역의 골목길에서 우리는 담장에 식물을 심은 화분들이 놓여 있고 벽의 틈새 공간에서 식생들이 자라나는 것을 종종 볼 수 있다. 평범한 담장에 식생의 레이어가 더해지면서 사유와 공공의 경계를 나누던 담은 그 길을 다니는 사람들을 위한 공공의 경험으로 환원된다.

리빙 프로젝트는 이러한 일련의 관찰과 경험에 기반하여 건축적으로 접근할 수 있는 가능성을 탐구한 프로젝트이다. 모듈에 기반한 단순하지만 건축적인 디자인 접근 방식을 통해 일상의 풍경을 건축화하고자 했다. 본 프로젝트는 화분 및 단위 벽돌에서 스트리트 퍼니처 시스템에 이르기까지 다양한 스케일에 걸쳐 적용되었다.

리빙브릭(Living Brick)

측면에 식생이 가능한 주머니를 가진 돌출형 벽돌을 제안한다. 돌출된 틈새는 벽돌 표면에 자라는 식생에 공간을 제공하며, 일상의 구축 행위와 식생 환경을 건축 외피의 일부로 자연스럽게 통합했다. 기성 벽돌과 함께 사용이 가능하도록 시공의 보편성을 확보하였다.

설계	남정민
설계담당	주병규
용도	건축 재료
크기	230×90×57mm(L×W×H)
재료	GFRC
제작	KSS
설계기간	2016. 11 – 2017. 1
사진	노기훈

유닛 다이어그램
Unit Diagram

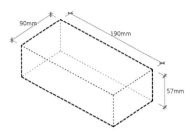

Typical Brick size 190×90×57mm

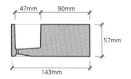

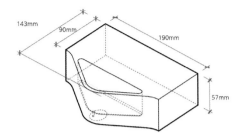

Living Brick size 190×90×57mm

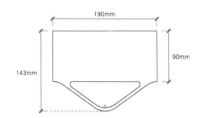

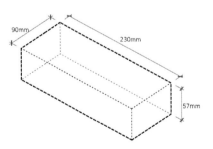

Typical Brick size 230×90×57mm

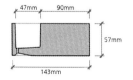

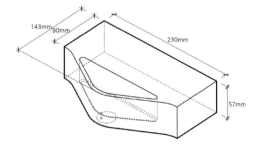

Living Brick size 190×90×57mm

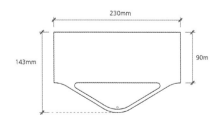

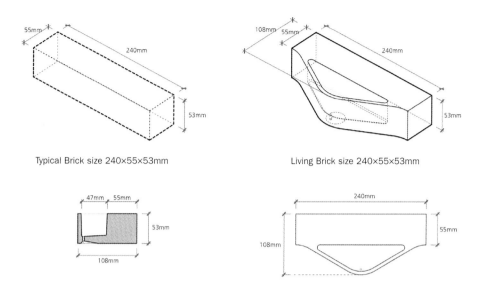

Typical Brick size 240×55×53mm

Living Brick size 240×55×53mm

단면 다이어그램
Section Diagram

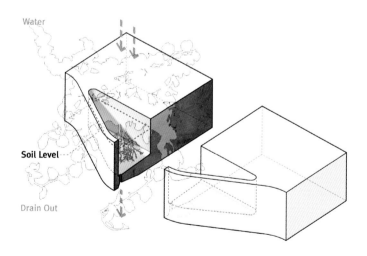

적층 다이어그램
Process Diagram

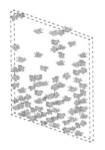 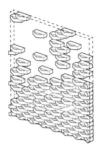 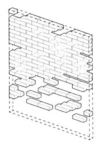 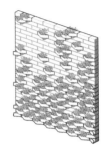

남정민

리빙블록(Living Block)

리빙블록은 단위 모듈의 유닛에 복합적인 다면의
틈새 공간을 적용한 프로젝트이다. 하나의 유닛으로
조합해 복합적인 패턴과 형태 구성이 가능하도록
계획했다. 리빙블록의 각 면은 각기 다른 모습의
정면성을 띠며 단독으로는 다면체의 화분이면서
동시에 식생을 위한 깊이를 가지는 구축용 단위
블록이 된다.

설계	남정민
설계담당	김병수
용도	건축 재료
크기	200×200×200mm(L×W×H)
재료	GFRC
특허	한국 제10-1701001호(2017. 1. 23)
	미국 US 9,901,036 B2(2018. 2. 27)
제작	시험 모델 자체 제작
설계기간	2014. 4 – 2015. 1
사진	노기훈

한 가지 모듈 = 여섯가지 면
One Module, 6 Sides

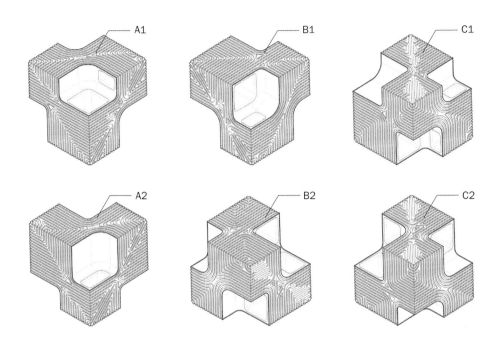

단면 다이어그램
Section Diagram

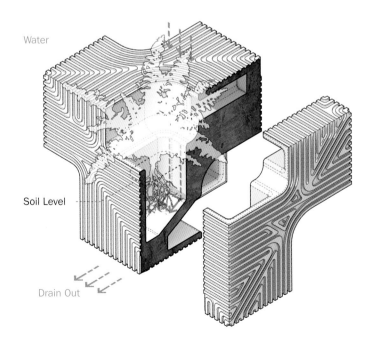

모듈 다이어그램
Module Diagram

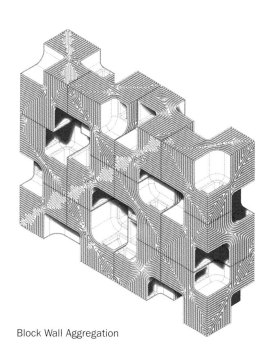

Block Wall Aggregation

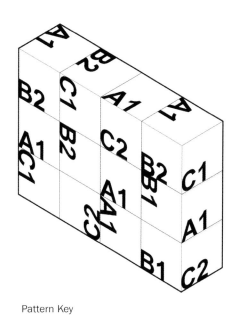

Pattern Key

패턴 전개도
Unfolded Pattern

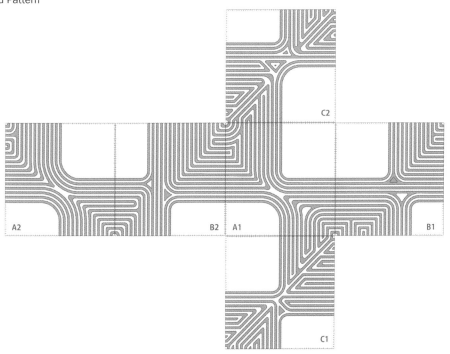

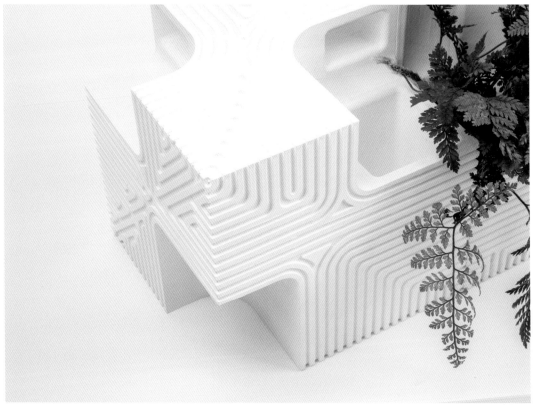

리빙포켓(Living Pocket)

리빙포켓은 얇은 입면에 틈새를 만들어 식물이
자랄 수 있는 공간을 제공한다. 이를 통해 기존 외피의
기능을 식생 영역까지 확장한다. 리빙포켓은 기존의
전형적인 돌 외벽 시공법을 적용할 수 있도록 패널로
디자인되었다.

설계	남정민
설계담당	김병수
용도	건축 재료
크기	600×600×150mm(L×W×D)
재료	GFRC
제작	KSS
설계기간	2013. 11 – 2014. 7
사진	노기훈

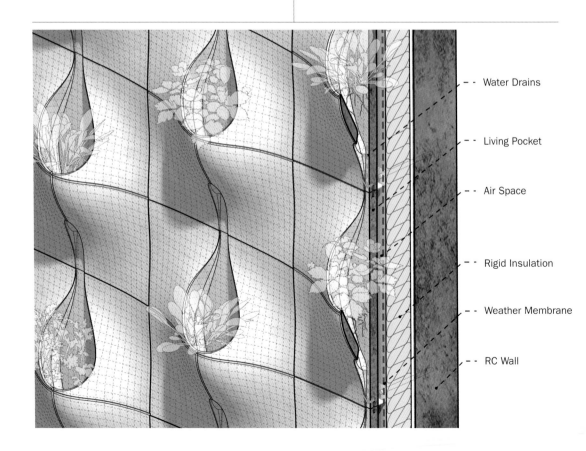

Water Drains

Living Pocket

Air Space

Rigid Insulation

Weather Membrane

RC Wall

리빙 프로젝트

리빙홀(Living Hole)

리빙홀은 단위 모듈에 대한 조합과 깊이에 대한
고민을 바닥으로 확장한 프로젝트이다. 기본적으로
구멍과 형상이 다른 세 종류의 유닛으로 구성되며,
이들의 조합을 통해 다공성 표피를 수평적으로
형성한다. 자연발생적으로 잡초가 자라나는 도시의
보도블록 틈새처럼, 콘크리트로 만들어진 세 종류의
유닛은 다른 형태와 크기의 작은 홈을 가지며 작은
자연을 담는 매체가 된다.

설계	남정민
설계담당	주병규
위치	경기도 화성시 효행로707번길 30 소다미술관
용도	건축 재료
크기	260×112×110mm(L×W×H)
재료	GFRC
제작	KSS
설계기간	2018. 2 – 2018. 4
사진	노기훈

유닛 타입
Unit Type

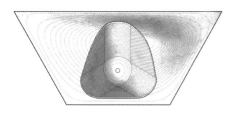

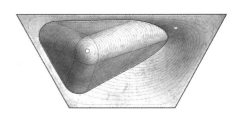

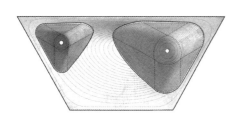

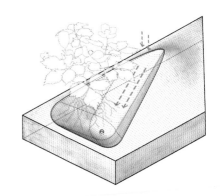

그룹 엑소노메트릭
Group Axonometric

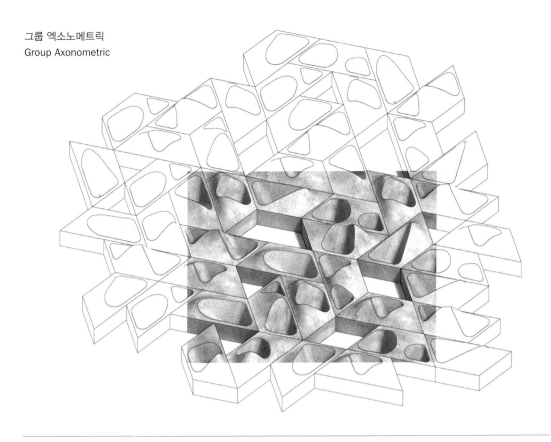

그룹 단면도
Group Section

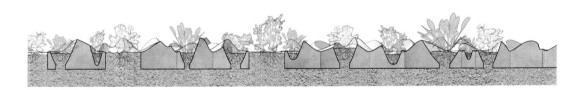

남정민

리빙퍼즐(Living Puzzle)

리빙퍼즐은 모듈에 기반한 식생 블록의 건축적 접근을 스트리트 퍼니처 규모로 확장해 적용한 사례이다. 모듈은 정삼각형, 평행사변형, 사다리꼴로 이루어진 세 가지 타입이다. 계산된 크기와 높이 변화를 통해 서로 조합되고 연결되면서 3개의 유닛만으로 다양한 형태의 스트리트 퍼니처로 재구성이 가능하다. 높이 변화에 따라 자전거 거치, 휴식 등의 기능을 담당하며 동시에 강변의 식생 환경으로 작용한다.

유닛 다이어그램
Unit Diagram

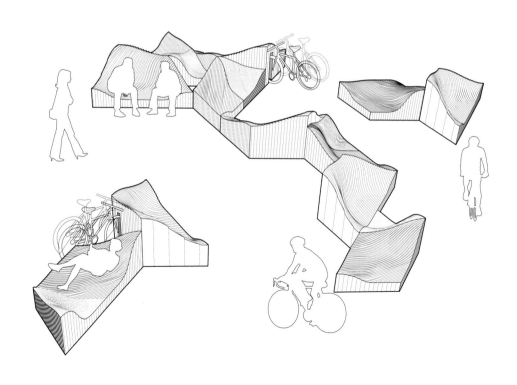

사이트 엑소노메트릭
Site Axonometric

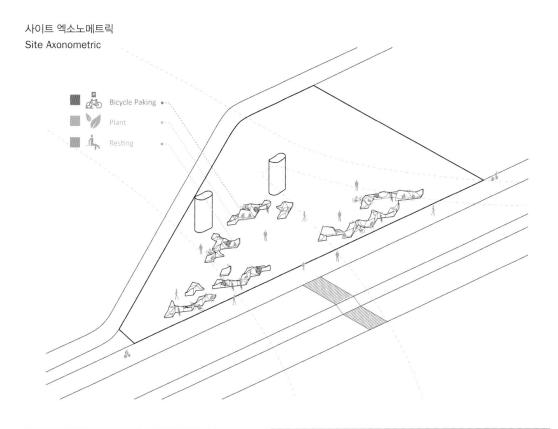

단면 다이어그램
Section Diagram

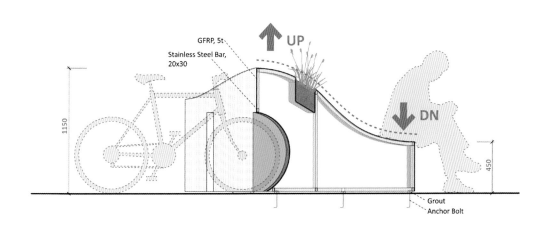

입면 다이어그램
Elevation Diagram

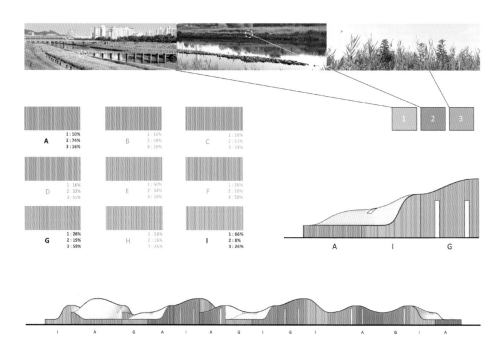

©Jungmin Nam

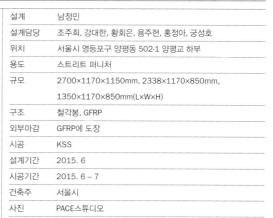

설계	남정민
설계담당	조주희, 강대한, 황회은, 용주현, 홍정아, 궁성호
위치	서울시 영등포구 양평동 502-1 양평교 하부
용도	스트리트 퍼니처
규모	2700×1170×1150mm, 2338×1170×850mm, 1350×1170×850mm(L×W×H)
구조	철각봉, GFRP
외부마감	GFRP에 도장
시공	KSS
설계기간	2015. 6
시공기간	2015. 6 – 7
건축주	서울시
사진	PACE스튜디오

작은공원
Alley House

작은공원은 반포동에 위치한 다가구주택과 근린생활
시설 프로젝트다. 서울의 급속한 도시화 과정에서
자연은 종종 우선순위에서 밀려 소외되어 왔다. 특히
다양한 욕망이 한데 뒤섞인 이곳 근린생활 밀집지역은
자연 소외현상이 가장 극심하게 나타나고 있다.
이 지역은 최소 법적 조경 면적조차 적용되지 않는
규모로 건물들 간에 법적 최소 간격만 유지하며 도시
조직을 형성하고 있다.

이 동네를 거닐며 유심히 관찰해보면, 건축가 혹은
공공의 손길이 닿지 않는 척박한 곳에서도
골목골목마다 건물 사이의 틈새에서 녹색의 생명이
피어나고 있다. 지극히 자연발생적인 녹화와 거주자의
녹색에 대한 욕망이 만들어낸 패턴의 조경이 그
나름의 질서를 가지며 척박한 환경의 틈새에서
피어나고 있다.

작은공원 프로젝트는 골목에서 자생적으로 생겨나는
녹화에 대한 관찰을 바탕으로 건물의 주변부와 공용
공간에 휴식과 자연을 위한 틈새를 제공한다. 기존의
지역적 한계를 극복하며 이 지역의 거주환경에 대한
고민과 함께 근린생활 시설에서도 소박하게나마
일상에서 최대한 자연을 접하도록 대안을 담았다.

8m 도로에서 꺾이며 들어가는 주 진입 쪽 입면은
이 건물의 새로운 얼굴 역할을 한다. 붉은 고벽돌과
청고벽돌 및 리빙브릭(Living Brick)으로 구성된
주 진입 면은 8m 도로를 향해 살짝 사선을 그리며
만들어낸 깊이와 틈새로 식생의 공간을 만들어낸다.
표면의 깊이와 골목과 마주한 식생 공간들은
사유화된 영역의 입면이자 조경이면서 동시에
골목에 생기를 불어넣는 공공의 환경으로 환원된다.

주 진입 골목은 건물 외부의 열린 계단을 통해 건물
안으로 수직적으로 이어진다. 이 계단실은 이동만이
아닌 원룸이라는 척박한 주거 여건 속에서 사람들에게
작은 공원이 된다. 계단과 계단참의 폭을 조정해
만들어낸 틈새 공간에는 식생이 자랄 수 있는 공간이
확보되고 사람들이 휴식할 수 있는 공용 공간이
만들어진다. 건축가 없이 조성한 자연발생적 식생
패턴을 바탕으로 제안한 이 접근법은 척박한 여건의
계단에서 가능한 대안을 제시한다. 이를 통해 어둡고
접근이 불편했던 골목은 사람을 반기는 골목이 되며
보다 개선된 도시 조직의 일부가 된다.

작은공원

남정민

©Yousub Song

남정민

작은공원

©Yousub Song

©Yousub Song

남정민

작은공원

©Yousub Song

설계	남정민
설계담당	임홍량, 주병규
위치	서울시 서초구 반포동 725-13
용도	다가구 및 근린생활
대지면적	135.9㎡
건축면적	81.22㎡
연면적	260.01㎡
규모	4층
높이	14.06m
주차	4대
건폐율	59.76%
용적률	191.32%
구조	철근콘크리트조
외부마감	벽돌, GFRC 블록
내부마감	나무, 석고보드위 도배, 타일
구조설계	건축사사무소 대룡
기계설계	(주)대오엔지니어링
전기설계	(주)대오엔지니어링
설계기간	2016. 7 – 2017. 1
시공기간	2016. 10 – 2017. 7
건축주	정규태
사진가	신경섭

매스 다이어그램
Mass Diagram

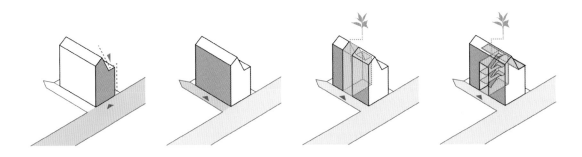

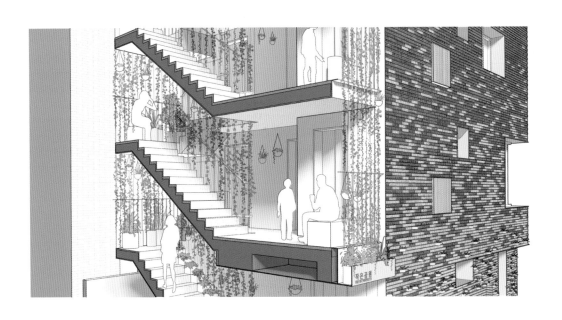

단면 투시도
Section Perspective

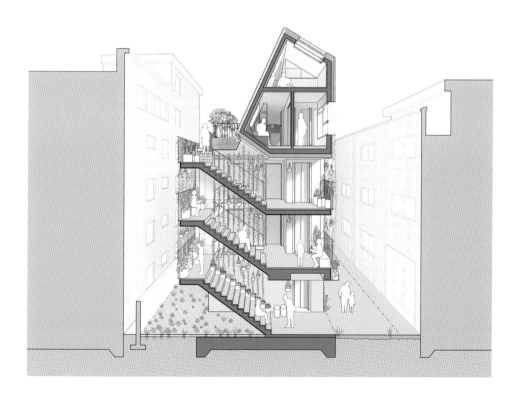

평면도
Plan

3F

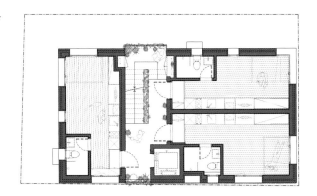

2F

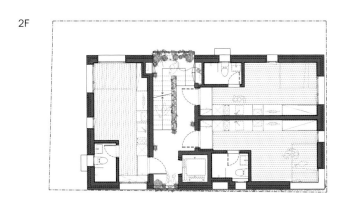

1F

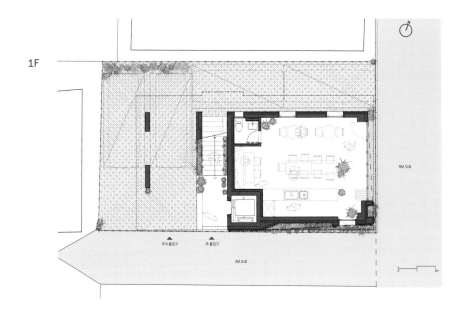

Roof

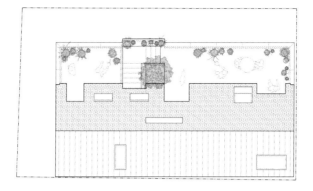

Attic

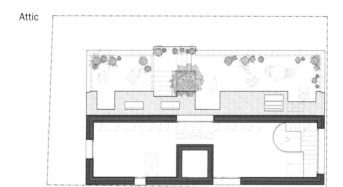

4F

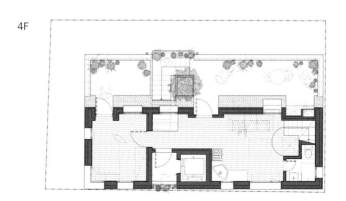

단면도
Section

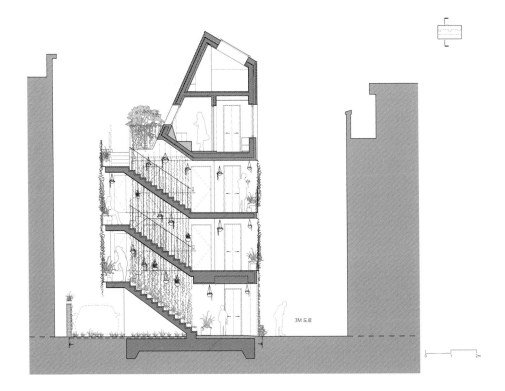

3M 도로

입면도
Elevation

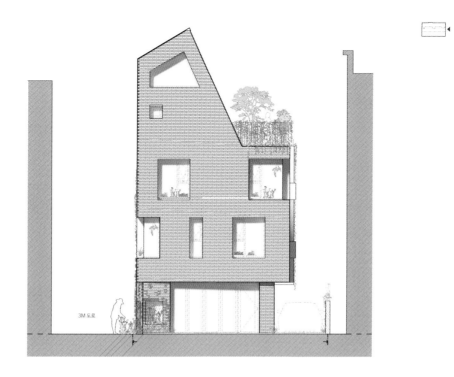

3M 도로

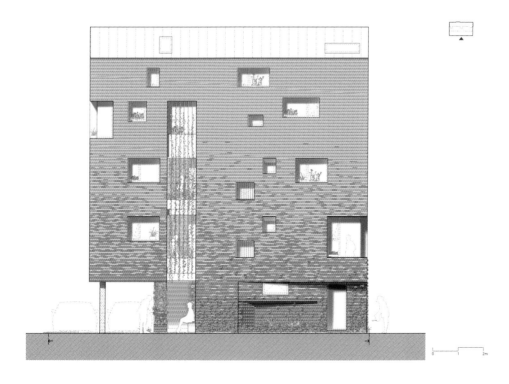

0　1　　　　2m

옐로우 풋
Yellow Foot

옐로우 풋은 서초동에 위치한 나홀로 아파트 프로젝트로, 인근에 다세대와 근린생활 시설이 밀집해 있다. 이 지역에서는 건물과 거리 간의 소통을 찾아보기 힘들다. 건물들은 담으로 둘러져 있거나 도로를 고려하지 않은 저층부 때문에 작은 섬처럼 이웃과 길과 분절되어 있다. 공원 같은 공공시설도 부족해 인근 작은 놀이터가 동네의 유일한 공용 공간의 역할을 하지만 이 지역을 지원하기에는 부족하다. 옐로우 풋 아파트는 이런 여건 속에서 길과 건물이 보다 열리고 소통 가능한 관계를 맺도록 시도하였다.

주차 공간 및 용적률 확보를 위해 적용된 필로티는 손가락으로 깍지 끼듯 지상 진입부 영역의 표면적을 늘리며 조경과 함께 도로를 향해 뻗어나간다. 이를 통해 차량 및 보행자의 진입부를 자연스럽게 형성해 동네를 향해 열린 입구를 조성했다. 이웃 건물들과는 달리 조경을 구성하는 자연이 도로변을 향해 관입을 시도함으로써 도로변에 열린 경계를 만들어 이 지역의 삭막한 풍경에 활기를 불어넣는다.

건물의 입면은 거실과 방에 따라 두 종류 크기의 창으로 이루어진다. 외부 마감재인 화강석의 간격에 따라 각 유닛별로 규칙을 갖고 엇갈리는 창은 규칙적 입면에 임의적인 패턴을 부여한다. 여기서 30cm 폭으로 움직이는 창을 고려해 내부 평면에서도 변동 가능 구간과 고정 구간으로 외벽 주변의 영역을 구분해 외부 패턴이 내부와 함께하도록 계획되었다. 잔다듬 마감의 화강석은 홈파기의 방식에 따라 총 세 종류의 패턴을 가지며, 이를 통한 세 종류 음영의 조합과 돌출된 창틀의 세 가지 색상 조합은 건물에 풍부한 표정을 불어넣는다.

북쪽 일조에 따른 사선제한과 서초구의 아파트 테라스 자체 규정으로, 건물은 크기가 다른 2개의 명확한 매스가 위아래로 쌓이며 형성된다. 나홀로 아파트 개발 프로젝트로 진행된 한계 속에서도 옐로우 풋 아파트는 유닛 구성의 다양화를 꾀했다. 매스의 크기와 차량 및 보행자 진입부에 따라 중앙에서 한쪽으로 치우친 코어를 통해 2베드와 3베드 유닛에 기반한 총 6개의 다른 유닛을 가진다.

옐로우 풋 아파트는 길을 향해 닫힌 이웃 건물들 사이에서 홀로 열린 경계를 제공하며 다세대 및 근린생활 밀집지역에서 다뤄지지 못했던 길과 건물 간의 관계에 대안을 제시한다. 지상부는 아파트의 사적 공간과 공공의 길 사이에서 물리적·시각적으로 열린 경계를 제공하며 길을 걷는 사람들의 경험으로 공유된다. 여기에 길을 향해 입체감을 가지며 서 있는 상부의 입면이 더해진다. 옐로우 풋 아파트를 통해서 개발 위주의 주변 지역에 긍정적인 변화가 일어나길 기대한다.

남정민

옐로우 풋

남정민

©Yousub Song

설계	남정민
설계담당	임홍량, 주병규
위치	서울시 서초구 서초동 1344-3~6
용도	공동주택
대지면적	790.9m2
건축면적	343.35㎡
연면적	2,591.63㎡
규모	지상 10층, 지하 1층
높이	32.2m
주차	31대
건폐율	43.41%
용적률	249.51%
구조	철근콘크리트
외부마감	화강석, 고흥석, 금속, 도장
내부마감	나무, 석고보드 위 페인트, 타일
구조설계	퀀텀엔지니어링
기계설계	천일 ENG
전기설계	천일 ENG
시공	(주)예미종합건설
설계기간	2015. 7 – 2016. 7
시공기간	2016. 1 – 2017. 4
건축주	(주)프레스티지
사진	신경섭

엑소노메트릭
Axonometric

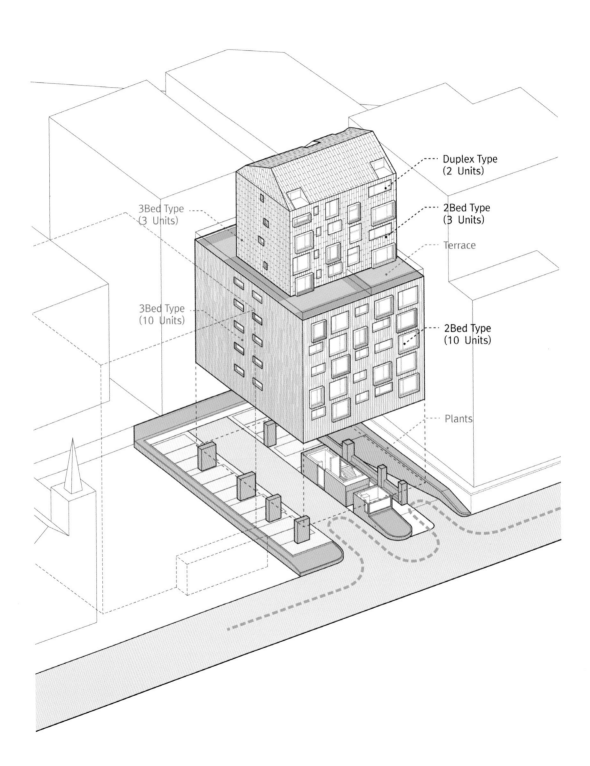

Duplex Type
(2 Units)

2Bed Type
(3 Units)

Terrace

3Bed Type
(3 Units)

3Bed Type
(10 Units)

2Bed Type
(10 Units)

Plants

파사드 패널 디테일
Facade Panel Detail

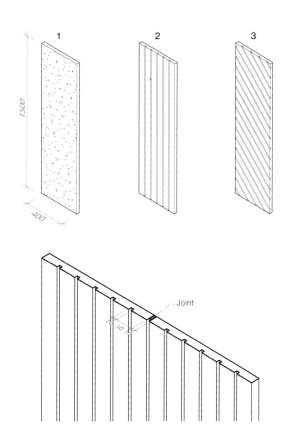

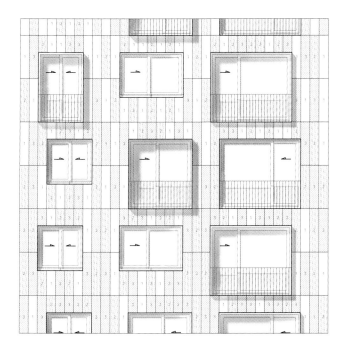

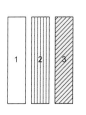

평면도
Plan

1F

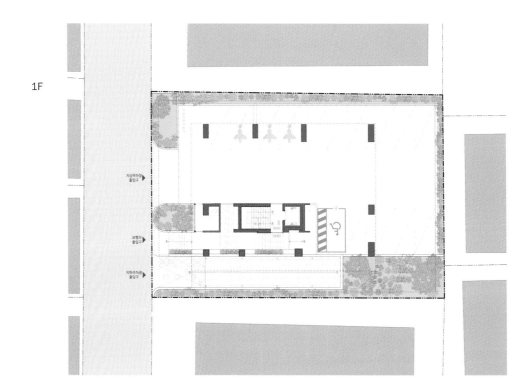

B1

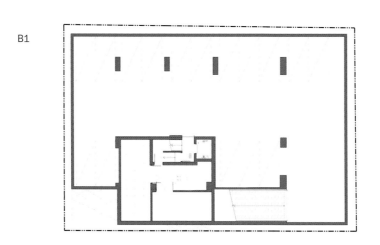

10F

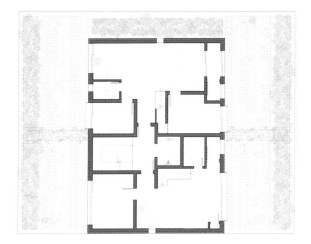

7–9F

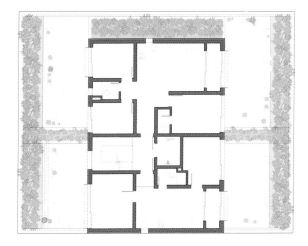

2–6F

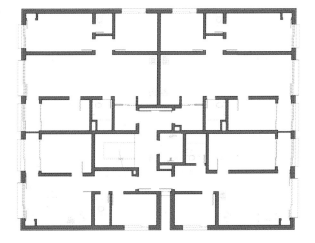

옐로우 풋

단면도
Section

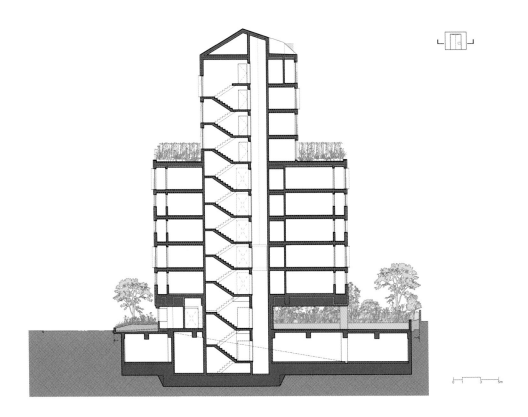

입면도
Elevation

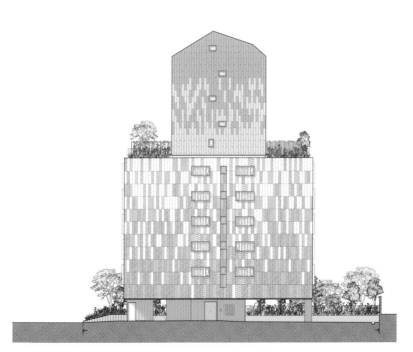

꽃+유치원
Flower+Kindergarten

꽃+유치원은 원생 수와 교실 면적을 최대한 확보해야 하는 도심의 여타 유치원과 유사한 조건을 안고 시작했다. 프로젝트의 핵심은 이를 충족하면서도 기존의 중복도 형식의 획일적인 배치에서 벗어날 수 있는 건축적 해법이었다. 용적률의 가치와 교육 공간으로서의 가치를 조율해야 하는 상황. 용적률은 최대한 확보하되 공간적 풍부함을 통해 어린이들의 창의성을 자극하고 자연을 접하는 유치원. 계절과 시간의 변화를 일상에서 느끼며 어린이들이 놀이와 배움, 자연의 경험을 더욱 밀접하게 체험할 수 있는 유치원을 계획하려 했다.

표면에 깊이를 주는 방식으로 그 해법을 찾았다. 건물 외피에 깊이를 주어 식생의 틈새와 창의 깊이를 만들어내고, 외부와 접하는 내부 산책로를 형성하여 건물 내외부의 관계를 확장했다. 내부 산책로는 나선형의 계단으로 I층에서부터 최상층까지 층별로 다른 위치에서 홀과 만나며 층과 층 사이를 입체적으로 연결한다. 이곳은 미끄럼틀과 계단, 틈새 공간으로 이루어진 하나의 거대한 놀이터를 형성한다. 어린이들은 오르내리는 일상의 행위에서 공간적 풍부함과 시각적 자극을 받고, 주변의 풍경과 계절을 360도의 파노라마로 경험한다.

각 층마다 3개의 교실과 하나의 다목적 홀을 배치해 복도가 없는 유치원을 만들었다. 다목적 홀은 제2의 교실이자 놀이 공간으로서 다양한 기능과 동선을 수용한다. 홀을 둘러싼 교실들은 구조에서 자유로운 내벽으로 구획되고 내벽은 곡면을 그리며 입체적인 교실의 형상을 만들어낸다. 이를 통해 교실은 형태적으로 개별성을 갖고, 두께를 갖는 내벽에 수납 및 놀이가 가능한 다양한 기능의 틈새 공간이 만들어졌다.

내부는 층별로 다른 컬러가 지정되어 층수 대신 노랑, 핑크 등의 색상을 통해 공간을 인지할 수 있다. 동일한 계열 안에서 건축 요소에 따라 다른 색조가 적용되어 색상 간의 미묘한 차이를 일찍부터 인지할 수 있는 환경을 조성하였다. 내부의 색상이 하얀 건물의 계단실 틈새로 흘러나오며 건물의 외부 입면에 새로운 표정과 이미지를 만들어준다.

대지 형태를 따르는 매스와 연속된 띠창, 각 교실마다 어린이와 어른의 스케일에 맞추어 정교하게 배치된 크고 작은 창, 그리고 그 사이로 새어 나오는 색상을 통해 꽃+유치원은 아파트단지의 반복적이고 분열적인 상황 속에서 역설적으로 작지만 새로운 랜드마크가 되어 단지 안 길잡이가 되었다.

꽃+유치원

남정민

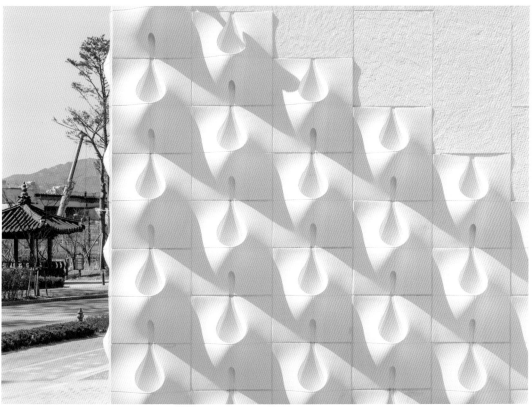

층별 다이어그램
Exploded Diagram

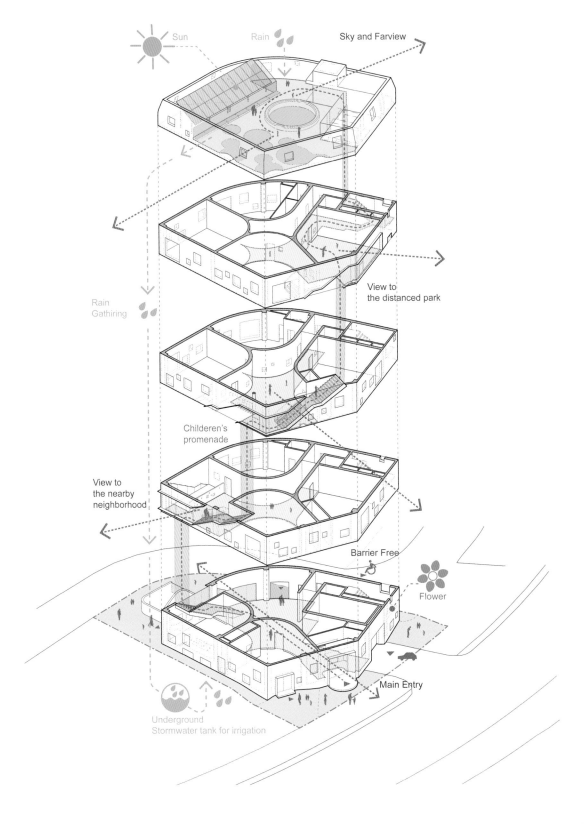

Sun

Rain

Sky and Farview

View to
the distanced park

Rain
Gathiring

Childeren's
promenade

View to
the nearby
neighborhood

Barrier Free

Flower

Main Entry

Underground
Stormwater tank for irrigation

남정민

설계	남정민
설계담당	김병수, 서정수
위치	서울시 서초구 우면동 731번지
용도	교육시설
대지면적	608m²
건축면적	303.44m²
연면적	2165.36m²
규모	지상 4층, 지하 2층
높이	18.5m
주차	8대
건폐율	49.91%
용적률	199.61%
구조	철근콘크리트조
외부마감	크레마 빌라, GFRC 패널
내부마감	나무, 석고보드 위 페인트
구조설계	터구조
기계설계	보우 ENG
전기설계	보우 ENG
시공	예미건설
설계기간	2013. 5 – 2015. 1
시공기간	2013. 10 – 2015. 1
건축주	예원유치원
사진	신경섭

컬러 계획
Color Application

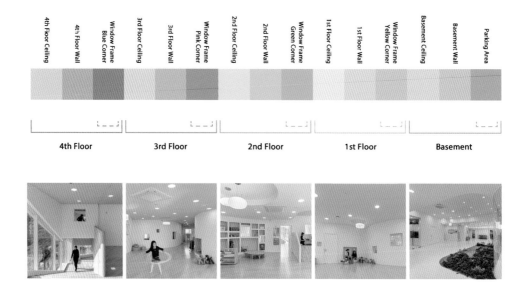

| 4th Floor Ceiling | 4th Floor Wall | Window Frame Blue Corner | 3rd Floor Ceiling | 3rd Floor Wall | Window Frame Pink Corner | 2nd Floor Ceiling | 2nd Floor Wall | Window Frame Green Corner | 1st Floor Ceiling | 1st Floor Wall | Window Frame Yellow Corner | Basement Ceiling | Basement Wall | Parking Area |

4th Floor   3rd Floor   2nd Floor   1st Floor   Basement

입면별 조경계획
Unfolded Elevation and Planting Patterns

Green 81 Modules / Pink Flower 3 Modules   Pink Flower 73 Modules   Purple Flower 19 Modules   Yellow Flower 87 Modules

구획 다이어그램
4 Division Multi Hall

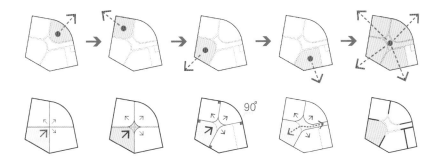

단면도
Section

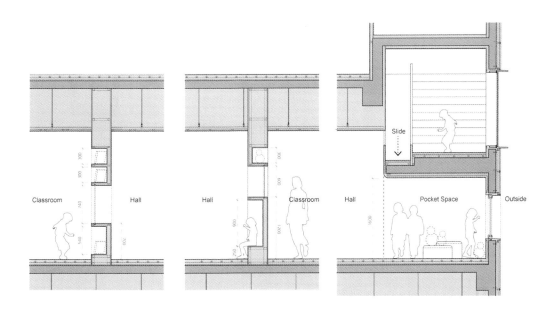

평면도
Plan

1F

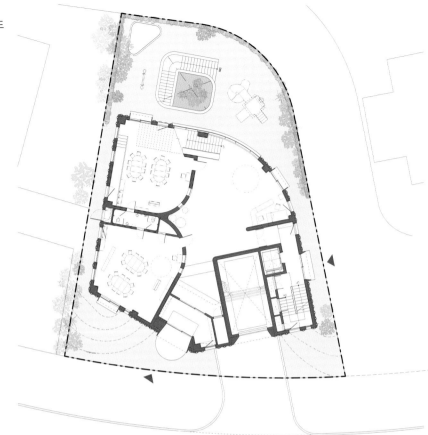

B1

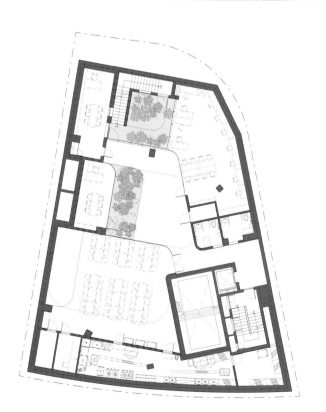

N

0 1   3   5m

남정민

4F

RF

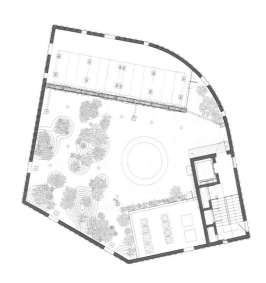

2F

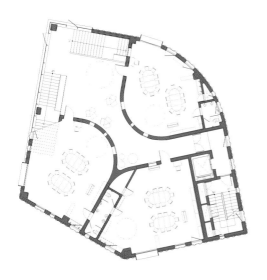

3F

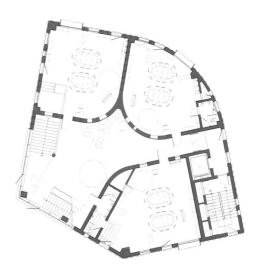

단면도
Section

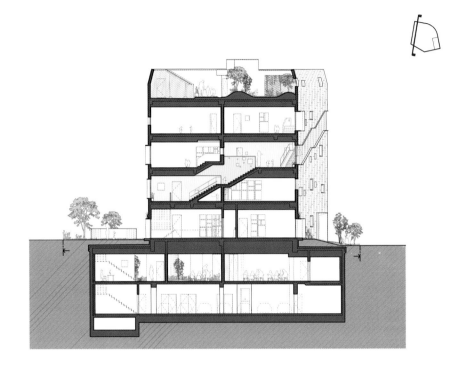

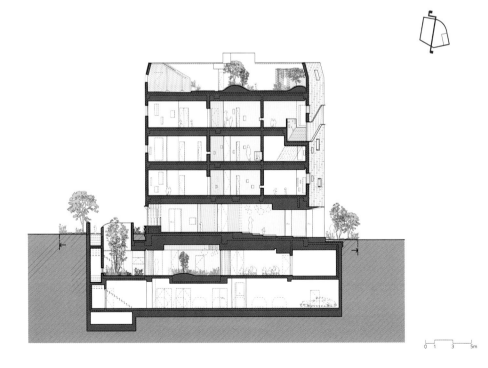

0 1 3 5m

입면도
Elevation

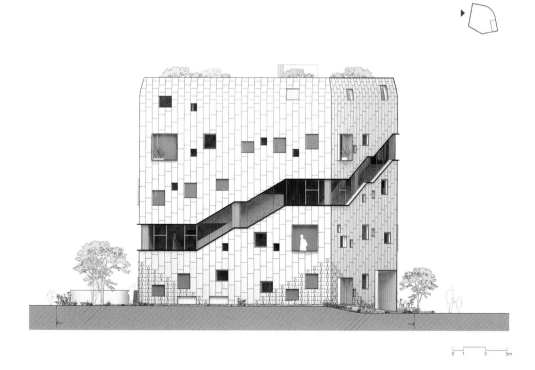

비평

표면의 깊이와 뿌리의 깊이
─젊은 건축가 남정민론

김현섭·고려대학교 교수

"표면은 건축물에 있어서 가장 다수에 의해서 공유되어 경험할 수 있는 요소이며 건물의 안과 밖의 다양한 영역에 걸쳐 우리의 일상과 밀접하게 맞닿아 있다. 그중에서도 건축물 외부의 표면은 건물과 도시의 영역 사이의 미묘한 접경지에서, 개인의 욕망과 도시의 공공성이 충돌하는 치열한 경계선에 놓인 첨병 역할을 하며 우리의 일상에 존재해왔다. 대부분의 건물들의 경우, 우리는 그 건물들의 표면을 경험한다. 개별 건물의 표면이 모여서 만들어낸 공용 공간(길, 광장 등)을 통해 우리는 도시의 공간을 경험하기에, 서울과 같은 개발 중심의 고밀도 도시에서 표면이 가진 건축의 공공적 가치는 역설적으로 더욱 커질 수 있다. 비록 건물은 개인의 욕망으로 지어지지만 표면을 통해서 공공과 대화를 하고, 표면이 2차원의 평면을 넘어선 깊이[를] 가질 때 더 풍부한 이야기를 할 수 있는 가능성이 생긴다. 사유와 공유 사이의 치열한 접경지에 놓여 있는 표면의 범위를 건물의 입면, 내부, 바닥, 도로변 경계로 확장하고, 표면에 적용 가능한 깊이에 대한 고찰을 통해 경계에서 발생하는 미세한 틈새가 가진 건축적 가능성을 탐구하고자 한다."[1]

— 남정민, 「표면의 깊이」, 2018

1) 2018년 젊은건축가상 지원을 위해 제출한 포트폴리오 「표면의 깊이: 건축물과 도시, 사유와 공공의 경계에서 발생하는 틈새 공간의 가능성」에서 인용

만약 건축이 표면에 관한 것이라면

남정민의 포트폴리오가 '표면의 깊이'를 테마로 내세운 것은 근래의
건축 경향에 대한 적절한 반영이라 할 만하다. 한동안 표면(surface)은
공간(space)이나 구조(structure) 등 건축의 핵심 이슈에 비해
부수적이고 하찮은 주제로 치부되어왔다. 하지만 따지고 보면 르
코르뷔지에의 '자유로운 입면'이든 로버트 벤투리의 '장식된 헛간'이든,
실상은 모두 표면에 대한 관심의 표명에 다름 아니다. 좀 더 가까운
시점으로 눈을 돌려 봐도 그렇지 않다. 우리 몸의 촉각이 건물 외피의
마티에르와 어찌 교감할 수 있는지에 대한 현상학적 해석이든 디지털
테크놀로지를 가능케 한 미디어 파사드의 유희든, 현대의 많은 건축적
이슈가 표면을 기반으로 하지 않느냐는 말이다. 이 같은 건축의 표면에
대한 관심은, 예컨대 데이비드 레더배로와 모센 모스타파비가 십수 년 전
『표면의 건축』(2002)에서 논의했던 바의 연장선상에 있다고 볼 수 있다.
이들은 건축 표면에 표출된 생산(production)과 표상(representation)
사이의 불일치에 주목하면서도, 그 가운데서 새로운 건축 실천과 담론의
가능성을 탐색했다. 리콜라 창고(Ricola Storage, 1987)나 에버스발데
도서관(Eberswalde Library, 1996) 등에서 헤르초크 앤드 드뫼롱(Herzog
& de Meuron)이 보여준 '표면 효과'가 그런 가능성의 대표적 사례라는
것이다.[2] 이리 볼 때 우리는 관점만 조금 달리한다면, 건축의 표면에서도
충분한 '깊이'를 찾을 수 있겠다.

이상은 젊은 건축가 남정민이 제시한 '표면의 깊이'에 대해 필자가
우선적으로 떠올린 바다. 특히 남정민이 유학했던 하버드대학교
GSD(Graduate School of Design)의 학장이 모스타파비라는 점을
감안하면, 그의 테마에 『표면의 건축』을 결부하는 것은 자연스럽다.
허나 필자의 질문에 남정민은 이 책을 최근에야 알았고 표면에 대한
관심은 이전부터 발전시켜온 것이라 답한다. 비록 주제에 대한 확신과
표면이라는 단어만큼은 이 책에서 왔음을 인정했지만 말이다.[3]
GSD에서는 오히려 그가 직접 수업을 들었던, 파시드 무사비나 앙투앙
피콩(Antoine Picon)의 저서가 표면에 대한 관심에 더 영감을 줬다는
것이다. FOA(Foreign Office Architects)의 창설자였던 무사비는
『장식의 기능』(2006)에서 특정 효과의 매개자로서 장식을 유형별로
분석했고,[4] 무사비에 고무되기도 했던 피콩은 『장식: 건축의 정치학과
주체성』(2013)에서 '장식의 귀환'이 내포한 새로운 의미의 가능성에
대해 논했다.[5] (20세기 초 아돌프 로스(Adolf Loos) 이래 제거됐던
장식이 돌아왔다!) 사실 남정민은 두 저서를 일부만 참조했다고 한다.
그러나 디지털 디자인 및 패브리케이션과 관련된 내용에는 흥미를
느꼈다는 것이다. 아마도 그런 테크놀로지가 주는 장식 혹은 표면 효과에

2) David Leatherbarrow and
Mohsen Mostafavi, *Surface
Architecture* (Cambridge
MA: MIT Press, 2002), pp.
209–214.

3) 필자와 남정민의 대화 및
이메일 교환, 2018. 7. 17–18

4) Farshid Moussavi and
Micael Kubo (ed.), *The
Function of Ornament*
(Barcelona: Actar, 2006)

5) Antoine Picon, *Ornament:
The Politics of Architecture
and Subjectivity* (Chichester:
Wiley, 2013)

고무됐으리라. 피콩의 저서는 다음 문장으로 시작한다. "만약 건축이
결국은 장식에 관한 것이라면 어찌할 것인가?" 그가 몇 단락 뒤 한 시인의
입을 빌려 "어떤 것도 사실상 표면보다 깊지 않다."고 적은 것을 보면, 이
문장은 다음과 같이 바꿔볼 수 있다. "만약 건축이 결국은 표면에 관한
것이라면 어찌할 것인가?"

표면의 실천

하지만 건축의 '결국'이 표면에 관한 것인지는 더 따져볼 문제이며,
건축에선 여전히 공간과 구조와 텍토닉 등의 이슈가 훨씬 핵심적이다.
그럼에도 최근의 건축계에서 표면의 중요성이 복권됐음은 틀림없다.
그리고 남정민은 그 같은 흐름 속에 있는 건축가다. 스스로는 더 큰
주제를 욕망하는 것으로 보이지만…….[6] 그렇다면 그의 실제 디자인은
표면을 어떻게 '실천'하고 있는가?

미국에서의 유학과 실무를 거쳐 2013년 여름 귀국한 그는, 지금까지
상당히 튼실하게 경력을 쌓아가고 있다. 사무실을 운영하던 중 그해
9월부터 서울과학기술대학교에서 교수로 일하기 시작했고, 곧 자기
건축연구소 OA-Lab을 열었으며, 2015년부터는 서울시 공공건축가로도
활동 중이다. 그러나 이런 공식적 직함보다 중요한 것이 어떤 디자인을
어떻게 실현했느냐에 있음은 두말할 나위 없다. 귀국 후 현재까지 5년
동안 그는 3개의 중규모 건축물, 즉 꽃+유치원(2013-2015), 서초동
아파트(2015-2017), 작은공원(2016-2017)을 실현했다. 서초구 우면동에
위치한 꽃+유치원은 그의 데뷔작으로서 이미 《공간》(2016. 2) 등에
소개됐고 몇몇 상을 받기도 했는데,[7] 벽면을 수놓은 꽃이 인상적이다.
그리고 서초동 아파트는 9층의 단일 블록에 28세대의 아파트를 담고
있으며, 1층 필로티 공간의 노란색 기둥 때문에 영어로는 'Yellow
Foot'이라 명명됐다. 한편 반포동의 작은공원은 이름과 달리 'Alley
House'라는 중소형 공동주택 건물이다. 1층은 상점, 2-3층은 도합
여섯 유닛의 원룸, 4층은 건축주의 주택으로 계획되었다. 남정민은
이 세 건축물에 더해 리빙 프로젝트(Living Project)라 명명한 각종
벽돌·블록 화분 및 스트리트 퍼니처를 선보인다. 그중 2015년 고안된
리빙블록(Living Block)은 최근 한국과 미국에서 특허를 얻은 독특한
화분 블록 유닛으로, 여러 개를 조합하면 일종의 벽을 구축할 수도 있다.
여기까지가 이번 젊은건축가상 공모에 제출된 작품들인데, 거기에는
서울시 공공건축가로서 실현한 송천동 주민센터 부분 리모델링(2017)과
몇몇 공공건축 계획안은 포함되지 않았다. 젊은 건축가 치고는
꽤 나쁘지 않은 실적이렷다.

6) '젊은' 그에게 '표면'이라는
주제는 아직 잠정적인 듯
보인다. 필자와 남정민의 대화
및 이메일 교환, 2018.7.17-7.18

7) 2015년 미국건축가협회
국제지역상(AIA International
Region Design Awards)에서
명예상(Honor Award for
Architecture)을 수상했고
2017년 서초건축상 우수상을
수상했다.

상기의 건물들을 보건대, 남정민 건축의 공통적 특징으로 바로 눈에
띄는 것은 밝은 색상의 사용이다. 특히 노란색은 그의 팔레트의
베이스임에 틀림없다. 처음 꽃+유치원이 완공됐을 때, 각 층을 장식한
파스텔 톤의 가지각색에 대해서는 유치원이어서 그러려니 했을 법하다.
그러나 서초동 아파트에서도 작은공원에서도 그리고 송천동 주민센터
리모델링에서도 샛노란색이 건물 곳곳에 칠해져 남정민 건축의
아이덴티티를 부여하고 있다. 헌데 색상의 적용은 그리 엄격하진 않으나
나름의 규칙을 따르는 것 같다. 수직의 외피를 수평으로 끊는 면과 외피
안쪽으로 셋백(setback)된 공간이 색상을 덧입은 주요 대상으로 보이기
때문이다. 서초동 아파트에서처럼 전자에는 창호 프레임의 안쪽 면이,
후자에는 1층 필로티 공간의 천장, 벽, 기둥 면이 전형적으로 속한다.
꽃+유치원에서야 층별로 다른 색상을 사용한 것에, 그래서 건물을
감아 올라가는 나선형 계단과 띠창으로 그 빛깔이 번져 나오는 것에
주안점이 있었다. 그런데 군데군데의 커다란 창과 출입구 프레임의
안쪽 면에 색을 입힌 것도 같은 맥락이다. 그리고 작은공원에도 서초동
아파트에서의 규칙이 느슨히 적용됐는데, 1층 필로티 공간에서 4개 층에
이어지는 계단실로 노란 색조가 확장됨이 특기할 만하다. 주차를 위한
필로티 공간이든 계단실이든, 모두 건물의 공용 공간임을 일단 기억하자.
남정민의 색상 적용 전체를 포괄해 비유하자면, 그는 살갗을 절개해
속살이 드러난 부분에 색을 입히는 경향을 보인다. 다시 말해, 남정민의
건축에서 색상을 덧입은 부분은 '표면의 깊이'를 가장 즉각적으로
느끼게 되는 부분이라 할 수 있다.

허면 건물 속살의 색상이라는 관점에서 포획한 두 가지 소주제, 즉 창호
프레임과 공용 공간을 바탕으로 논의를 좀 더 전개해보자. 우선, 창호
프레임. 꽃+유치원과 서초동 아파트에서처럼 남정민은 여러 창호 가운데
강조해야 할 것을 선택해 안쪽에 색을 입혔는데, 색을 입히기 전에 전제된
것이 바로 그 프레임을 돌출하는 행위였다. 이로써 상인방과 하인방을
포함한 창의 프레임은 충분한 깊이를 얻게 되며, 표면의 깊이라는 주제를
곧바로 환기한다. 돌출된 창호 프레임은 햇빛을 가리는 일종의 루버와
같은 실용적 역할을 하는데, 서초동 아파트의 경우 옆 세대로부터의
시선 차폐라는 최소한의 기능도 수반할 것이다. 게다가 상실된 발코니에
대한 보상심리도 은연중 작동한 것 아닐까. 하지만 본고의 관점에서 더
유의미한 것은 역시, 건물 표면에 깊이를 줌으로써 그러지 않았다면
밋밋했을 파사드에 풍부한 표정을 덧입혔다는 점이다. 밝아진 건물의
표정은 단지 그 건물만이 아닌 주변의 도시적 맥락과 연동되는 까닭에 그
뜻이 배가 되리라. 더욱이 얇은 두께의 금속판으로 날렵하게 만들어낸
서초동 아파트의 돌출 프레임은 깊어진 표면과 대비를 이룸으로써
더욱 신선한 느낌을 불러일으킨다. MVRDV가 암스테르담에 설계한

보조코 아파트(Wozoco Apartments, 1994-1997)의 과감한 실험성에야
비하기 어렵지만 말이다. 그런데 창호 면을 후퇴시키든 창호 프레임을
돌출시키든 그로써 표면에 깊이를 주는 방식은, 건물의 외피에 깊이를
소거하는 또 다른 실험적 방식과 반대편에 있음이 흥미롭다. 한국적
콘텍스트에서는 승효상의 여러 디자인, 예컨대 웰컴시티 사옥(1995-
2000)을 떠올릴 만하다. 그는 여기서 건물 매스의 코르텐 강판 외피와
무테창 표면을 평활하게 일치시킴으로써 외벽의 무게감을 상쇄하는
효과를 가져온 바 있다.

표면의 공공성과 표면적 공공성

한편 창호 프레임의 경우보다 더 깊이 건물의 속살을 들춰 보이는 것이
노란색 주조의 공용 공간이다. 전술했듯 서초동 아파트의 I층 필로티
공간과 작은공원의 계단실이 여기 속한다. 건물의 표면에 대한 논의를
이런 깊숙한 공간으로까지 확장하는 것에 다소간의 작위성이 개입된
듯도 하지만, (표)면과 (표)면이 만나면 결국 공간을 이루는 것일 테다.
앞의 깊은 창호 프레임 역시 이미 공간성을 내포했다. 머리글로 인용한
남정민의 문장은 표면의 집합이 구성하는 공간, 특히 '공용 공간'에
집중하고 있다. 그리고 이것은 결국 건축과 도시가 만나는 경계의
공공적 가치, 즉 공공성 문제에 이른다. 이는 그가 표면이라는 테마의
표면성에 주저하며 더 큰 담론을 욕망했던 결과일 것이다. 게다가 그는
공공건축가 아닌가.

먼저 우리는 공용(共用)이라는 말이 지시하는 두 가지 층위를 되짚어볼
필요가 있다. 이 말은 의당 사적인 용도가 아닌 함께 나눠 쓴다는 뜻을
담는다. 서초동 아파트와 작은공원 둘 다 공동주택인 까닭에 주차장과
계단과 복도의 공용 공간이 있는 것은 당연한 일이다. 이런 공간을
건물 거주자들이 서로 공유하고 공용하며, 거기서 공동체 의식까지
싹틔우기를 바란다. 이것이 공용의 첫 번째 켜다. 그런데 그런 공용이
내부의 견고한 켜를 넘어서 익명의 불특정 다수에게까지 확장될 수
있을까? 관리인이나 집배원 혹은 특정 세대를 방문한 친지 등과 같은
관계자에게야 그게 가능하겠지만, 사유재산제의 자본주의 사회에서
거리의 낯선 보행자에게까지 공용이 허용되기에는 어려움이 만만치
않을 것이다. 이 공용 공간에 대해 빗대어 말하자면, 작은공원의 '공원'이
'共園'이지 '公園'은 아닐 거란 얘기다. 이런 현실적 한계를 인정한 연후에
남정민의 건축이 말하는 표면의 공공성(公共性)을 읽어야 하겠다.
그 공공성은 때로 표면적(表面的)이기도 하다.

서초동 아파트의 I층 필로티 공간은 무엇보다 시각적으로 공공적이다.

기둥과 최소한의 코어 부분만을 빼고는 모두 오픈된 덕에 정면의
도로에서 후면 화단까지 시각적으로 관통한다. 좌우의 이웃 아파트 블록
역시 필로티 공간을 두고 있으나 얕은 담장이나 차폐물로 경계를 짓기
때문에 그런 개방감을 누릴 수 없다. 또한 남정민의 필로티 공간과 화단의
노란 색상이 주는 선명성은 보행자의 시선을 사로잡기에 충분하며, 건물
앞 화단에 조성한 정원은 삭막한 도시 공간에 싱그러움을 주는 '共園'이자
'公園'으로 역할을 한다. 이 같은 지상부의 밝고 신선함은 건물 자체의
표면의 깊이가 주는 역설적 경량감으로 더 힘을 받고 있다. 여기서 '역설적
경량감'이라 함은 전술했던 깊은 창호 프레임의 얇은 두께가 주는 느낌을
지칭하는데, 외장 화강석 표면의 가는 실선 홈파기 패턴도 그런 감각에
일조한다. 서초동 아파트는 완공 후 주변에 재미있는 현상을 유발했다.
건물 앞에 노란색 의류수거함이 놓이고, 도로 건너편 다세대주택의
담장이 노란색으로 칠해지는 등 노랑의 확산이 발생한 것이다. 이에
대해 남정민은 '지상부의 공용 공간이 길과 의사소통을 이루어낸 결과'라
여긴다.[8]

작은공원의 계단실은 필로티의 공용 공간이 수직으로 확장된 바다. 이
계단 길에는 도시의 가로 체계를 건물 내부로 끌어오려는 의도가 명백히
깔려 있다. 한국 현대건축을 꼼꼼히 읽어온 독자에게라면 이런 의도가
결코 낯설지 않다. 더구나 남정민이 이를 두고 '골목의 연장'이라 칭한
어법은[9] 문자 그대로 익히 들어 본 바 아닌가. 조성룡이 양재 287.3 (1990-
1992)의 계단을 주변 '길의 연장'이라 하고, 승효상이 수졸당 (1992-
1993)의 마당을 '도시의 길에 이어진 골목의 연장'이라 표현했던 데서
말이다.[10] 우리는 그런 건축 어휘가 공동주택으로 실현된 예도 여럿
찾을 수 있는데, 방철린의 다가구주택 스텝 (1994-1995)과 이일훈의
다세대주택 퇴계불이 (1996), 가가불이 (1995-1996) 등이 대표적이다. 특히
이일훈의 '채 나눔'이라는 개념은 언급할 필요가 있다.[11] 작더라도 채를
나눠 이웃이 교감할 수 있는 공간을 부여해야 한다는 이 개념은, 부러
계단실의 폭을 넓히고 계단참의 깊이를 늘리며 작은공원에 '틈새의 공유
공간'을 만들고자 했던 남정민의 생각과 맥이 닿기 때문이다. 외장벽돌을
반으로 갈라 씀으로써 증대한 건물 연면적이[12] 결과적으로는 공유 공간을
확보하느라 잃어버린 사유 공간을 확충해준 셈이다. 그가 1990년대 선배
건축가들의 이론과 실천을 인지했는지의 여부는 여기서 그리 중요치
않다. 전 세대 건축가들의 노력이 한국 건축계에 공공성과 공동성에 대한
가치를 각인했음이 분명하고, 현재의 건축가들은 그 기반 위에 일하고
있기 때문이다. 단 남정민의 차별성은 계단실에 확보된 틈새 공간을
녹색의 식생에 할애했다는 점에 있다.

8) 남정민, 「표면의 깊이」, 2018

9) 남정민, 「표면의 깊이」, 2018

10) 4·3그룹, 『이 시대 우리의
건축』(서울: 안그라픽스, 1992)

11) 이에 대해서는 필자의
졸고를 보시오. 김현섭,
「4·3그룹 건축의 스펙트럼과
비판적 모더니즘」『종이와
콘크리트: 한국 현대건축 운동
1987-1997』, 정다영·정성규 편
(서울: 국립현대미술관, 2017),
78-87쪽

12) 그는 정교한 조적을
위해 모든 외장벽돌을 길이
방향으로 반으로 잘라 썼는데,
건물 네 면 모두에서 너비를
약 5cm씩 넓히는 효과를
가져왔다.

건축과 식물

그렇다. 남정민 건축만의 차별성은 건축에 식물을 도입하려는 노력에 있다. 최소한 한국 건축계에서 그렇고, 그도 아니라면 한국의 젊은 건축가들 사이에서는 그렇다. 건물과 화단이 어느 정도 분리된 서초동 아파트에서는 이 특성이 두드러지지 않을지 모르나 꽃+유치원과 작은공원에서는 식물의 도입이 그 건축물의 성격에 결정적이다. 이는 표면의 깊이와 깊게 관련되지만, 굳이 표면을 논하지 않더라도 그 자체만으로 독립된 건축 수법이라 하겠다.

꽃+유치원을 보자. 앞서 필자가 '벽면을 수놓은 꽃이 인상적'이라 적었듯, 1층 외벽면을 장식한 꽃과 그 꽃을 담는 일련의 화분 패널이 이 건물의 가장 큰 특징이다. 유치원 이름이 여기서 온 것임은 두말할 나위 없으며, 이에 비하면 지하 1층의 실내 정원이나 옥상의 정원은 매우 부수적으로 보인다. 그는 벽면에 부착하는 이 화분 패널을 리빙포켓(Living Pocket)이라 칭했는데, 앞에 언급한 리빙 프로젝트는 여기서 시작되었다. 리빙포켓은 한 변이 60cm인 정사각형 판을 한 유닛으로 해 유려한 곡면으로 중앙의 배를 불리고, 그 안에 포켓 공간을 마련한 콘크리트 패널이다. 물론 그 포켓은 흙과 식물을 담기 위함인데, 일반 화분처럼 물이 빠지도록 포켓 아래 구멍을 냈다. 그리고 여러 유닛이 연결되었을 때 포켓의 상단과 구멍 아래 물길의 아귀가 서로 잘 맞도록 디자인했으며, 패널이 위아래로 반씩 엇갈렸을 때도 각 유닛이 상호 합치되도록 패널 중앙에 수평 홈을 냈다. 이와 같은 리빙포켓의 유닛들이 유치원 벽면에 연접해 부착되니, 물결 문양이랄까 일종의 주름이랄까, 매우 독특한 패턴을 형성했다. 그리고 그 패턴이 3차원 곡면인 까닭에 자연히 표면에 깊이를 주게 되는데,[13] 여기에 꽃과 식물이 담김으로써 그 깊이는 물리적 깊이를 넘어 정서적 깊이로까지 나아가게 된다. 무엇보다도 아이들이 잘 고안된 벽면에 꽃을 심고 가꾸는 행위는 정서 함양과 창의성 교육을 위한 이 유치원만의 특별활동일 것이다.

작은공원에도 리빙 프로젝트의 하나인 리빙브릭(Living Brick)이 적용되어 이 건물을 특징짓는다. 리빙브릭은 벽돌의 한쪽 면을 배불리고 그 안쪽을 비워 화분으로 사용한 것이다. 남정민은 건물 1층의 좌측면을 중심으로 리빙브릭을 집중 삽입하고, 2층 하단에도 이를 산발적으로 배치한다. 이로써 건물 외벽에 식물이 자랄 수 있는 장을 열었다. 이에 더해 정면과 좌측면의 외벽을 일부 후퇴시킨 '깊은 표면'에 작은 화단을 만들었는데, 모두가 함께 어우러져 말 그대로 '작은공원'을 형성한다. 한편 앞서 언급했던 계단실의 녹화 방식도 두드러진 요소다. 계단 사이의 틈새와 계단참에 화분을 놓거나 걸 수 있게 하고, 여러

13) 이와 유사한 선례는 여럿 있는데, 한국에서는 김찬중이 선보인 한강터널 벽면 디자인(2009)과 래미안 갤러리(2009) 등을 떠올릴 수 있다. 그가 단위 유닛을 디지털 툴로 디자인하고, 공장에서 (대량) 생산해, 현장에서 조립하는 방식은 남정민의 리빙 프로젝트에도 대체로 부합한다.

층의 난간을 수직으로 관통하는 일련의 와이어를 설치해 넝쿨식물이
그것을 타고 올라가게 만든 것이 그렇다. 또한 유별나진 않지만, 측면과
후면의 주차장과 정면 자투리땅의 보도블록 사이에도 자연스럽게
잡초가 자라도록 배려했다. 여기에는 별도로 고안된 블록이 쓰이진
않았으나 유사한 개념이 리빙홀(Living Hole)이라는 바닥 블록으로
발전한다. 이것은 현재 진행 중인 경기도 화성 소다미술관의 전시회
〈인공자연 Artificial Nature: 콘크리트 자연을 담다〉(2018.4–10)를 위한
출품작으로서, 야외용 바닥 블록 구멍에 식물이 자라도록 한 것이다.
이처럼 남정민이 자기 건물 이외의 용도로 디자인한 리빙 프로젝트로는
리빙퍼즐(Living Puzzle)이 대표적이다. 2015년 여름 영등포구의 양평교
아래 설치된 이 스트리트 퍼니처는 여러 개가 다양한 방식으로 맞춰질 수
있는 곡면 벤치이기도 한데, 자체 내에 자전거 거치대뿐 아니라 화분을
담아 식물과의 조화를 꿈꿨다.

건축에 식물을 도입한다는 남정민의 아이디어는 이미 그의 GSD
졸업설계인 '도시의 농장, 도시의 진원지(Urban Farm, Urban Epicenter,
2009)'에서도 중추적이었지만,[14] 사실 이게 그리 새로운 시도는 아니다.
환경문제로 생태계 회복에 대한 관심이 이전부터 건축과 도시 디자인에
갖가지 방식으로 표출되어왔기 때문이다. 예컨대 수직정원으로 유명세를
타고 있는 프랑스 식물학자 패트릭 블랑(Patrick Blanc)은 헤르초크
앤드 드뫼롱이나 장 누벨 등 여러 스타 건축가들과 협업하기도 했다.
한국에서도 다양한 시도가 있었는데, 지속성 측면에서 어려움이 많아
보이지만, 조민석의 신사동 앤 드뮐레미스터 숍(Ann Demeulemeester
Shop, 2007)이 세련된 벽면녹화로 눈길을 사로잡은 바 있다. 그런데
남정민에게서는 오히려, 일본의 건축사가이자 건축가인 후지모리
데루노부의 디자인이 더 강하게 연상된다. 공중에 띄운 차실(茶室) 등
매우 독특한 디자인으로 알려진 그 건축가 말이다. 그는 건축가 데뷔
초부터 건축에 식물을 도입하고, 석기시대 정도의 단순 기술을 대폭
활용한다는 원칙을 가지고 있었다. 벽과 지붕에 민들레를 심은 자택
'민들레의 집(タンポポ·ハウス, 1995)', 부추를 한 포기씩 화분에 담아
지붕에 줄지어 삽입한 '부추의 집'(ニラハウス, 1997), 풀로 지붕 전체를
덮고 꼭대기에 동백꽃 나무를 한 그루 심은 '동백꽃 성'(ツバキ城, 2000)
등을 기억하라.[15]

남정민에게서 그를 떠올리는 데는 몇 가지 이유가 있다. 첫째, 큰
틀에서 보면 후지모리가 건물의 공간이나 구조보다 '마감(仕上げ)'에
주안점을 둔 점이 남정민의 표면에 대한 관심과 유사하다. 둘째, 블랑
등이 보여준 벽면녹화에 비해 후지모리의 지붕식재는 테크놀로지의
개입이 상대적으로 덜하고 식물이 건축을 지나치게 압도하지 않는데,

14) 그는 여기서 도시
인프라스트럭처로서의 고층
수직농장 구조물을 제안하며
이것이 농작물 공급을 뛰어넘어
사회적, 문화적 변화의
진원지로서 역할하길 바랐다.

15) 후지모리 데루노부의
건축에 대해서는 필자가 여러
편의 논고를 발표한 바 있다.
대표적으로 다음 두 논문을
보시오. 김현섭, 「미래소년
'테루보'의 신나는 건축 모험」,
《공간》 536 (2012. 7), 16-21쪽.
Hyon-Sob Kim, 'The uncanny
side of the fairy tale: post-
apocalyptic symbolism
in Terunobu Fujimori's
architecture', *The Journal of
Architecture*, vol. 21, no. 1
(February 2016), pp. 90–117.

남정민의 경우도 마찬가지이다. 셋째, 후지모리와 남정민 모두 일상의
작은 거리에서 발견되는 친밀한 오브제에서 디자인의 영감을 받았다.
이에 대해서는 부연 설명이 필요할 듯하다. 남정민의 포트폴리오는
우리가 늘 마주하는 서울 골목길의 여러 풍경 사진에서 시작하는데,
거기에는 각종 화분의 식물과 넝쿨이 '자연발생적 식생'을 형성하고
있다. 그는 이런 식생이 평범한 길가의 건물에 입체적 표면을 부여한다고
보고, 건축가 없는 일상의 건축에서 자기 건축의 가능성을 찾는다고
적는다. 정도야 다르겠지만 그의 일상에 대한 천착은 후지모리가
'건축탐정단'과 '노상관찰학회'를 통해 도시의 골목길을 누비며 소소한
사물과 풍경을 관찰하고 기록했던 바와 연계될 만하다. 비록 후지모리는
이런 활동을 자기 디자인과 직결하는 데에 다소 유보적이었지만 말이다.
이 경우가 예시하듯, 건축에 식물을 도입하는 것과 관련해 둘 사이의
차이도 상당하다. 식물을 심고 가꾸는 방식 자체에서는 두 사람 모두
테크놀로지의 개입을 최소화했지만, 식물을 담는 유닛을 생산하는 데
있어서 남정민은 디지털 테크놀로지를 최대한 활용했다. 리빙포켓과
같은 곡면 유닛의 형틀을 제작하기 위해서는 이런 최신 기술의 사용이
필수적인데, 석기시대의 기술을 추구하는 후지모리에게는 전혀 어울리지
않는 방식이다. 살아온 시대와 교육 배경이 다르니 이런 차이는 오히려
자연스럽다. 한편 결과로 도출된 건축 속 식물이 무엇을 의미하는가에도
차이가 크다. 그 의미 방향성은 오히려 테크놀로지의 활용 정도에
역행하는 듯하다. 남정민에게 꽃과 식물이 소박한 일상을 지시한다면,
후지모리의 그것은 문명사 저변에 깔린 기술문명의 폐허를 상징하기
때문이다.[16)]

결어: 뿌리의 깊이 ─────

이처럼 후지모리에 대한 단상은 남정민 건축의 식물 도입이 앞으로
어떻게 의미의 지평을 넓혀갈 수 있을지에 새로운 실마리를 제시한다.
이가라시 다로가 편집한 『건축과 식물』(2008) 역시 그럴 것이다. 이 책은
일본적 맥락에서 건축과 식물에 대한 다각적 논의를 담았는데, 후지모리
챕터도 하나 삽입됐다.[17)] 남정민은 후지모리를 전혀 알지도 의식하지도
못했으나 필자와의 대화 후 뒤돌아보고는, GSD 졸업설계를 위한 자료
속에서 그의 '부추의 집' 사진을 발견한다. 그리 보면 학창시절 남정민의
'식물성'에 후지모리가 전혀 무관했던 것도 아니다.

헌데 마지막으로 기억해야 할 바는, (앞의 조민석도 그렇듯) 후지모리든
남정민이든 건축으로의 식물 도입이 완전히 성공적이지는 않았다는
사실이다. 혹은 여전히 실험 중이라 바꿔 말할 수도 있겠다. 후지모리의
지붕에 식재된 민들레와 부추가 이듬해 곧 말라버렸던 것처럼, 남정민이

16) Hyon-Sob Kim, op. cit.

17) 五十嵐太郎 編,
『建築と植物』(京都: INAX,
2008).

리빙포켓과 리빙브릭에 심은 꽃과 식물도 대부분 얼마 안 가 화사함을 잃고 시들어버린다. 두 사람 모두의 경우에서 의도하지 않았던 잡초는 잘 자라고 있고, 그럼에도 남정민이 작은공원의 넝쿨식물에서 희망을 보고 있다는 점을 생각하면, 어떤 식물을 선택하느냐가 중요할 테다. 하지만 그보다 더 중요한 관건은, 너무도 당연한 바, 식물이 뿌리를 내릴 수 있는 충분한 토양과 깊이를 마련하는 데 있다. 그래야 식물이 자생해 유지관리가 용이하고, 지속가능성을 담보할 수 있기 때문이다. 이 점은 결국 본고의 전체 주제와도 연통한다. 뿌리의 깊이를 위해 '표면의 깊이'를 더해야 할 것이다. 그리고 그만큼이나, 표면에 대한 사유 역시 건축역사와 담론의 물줄기에 깊이 뿌리내려야 할 것이다. 그럴 때에라야 표면의 사유는 '생산'과 '표상'의 불일치를 관통하든, 소박한 일상과 문명사 저변의 사이를 진동하든, 실한 열매로 맺히지 않겠나. 남정민은 아직 젊다. 고로 앞으로가 더 기대된다. 그의 사유의 뿌리가 더 깊이 뻗어나갔을 때, 그때 '중견 건축가' 남정민론을 다시 써보자. "불휘 기픈 남간 바라매 아니 뮐째 곶 됴코 여름 하나니……."

김현섭은 영국 셰필드대학교에서 서양 근대건축을 공부했고, 2008년 모교인 고려대학교 건축학과에 임용된 이래 건축역사·이론·비평의 교육과 연구에 임하고 있다. 근래에는 한국 현대건축에 대한 비판적 역사 서술에 관심을 모으고 있다. 최근 『건축수업: 서양 근대건축사』(2016), 『건축을 사유하다: 건축이론 입문』(2017), 「DDP Controversy and the Dilemma of H-Sang Seung's 'Landscript'」(JAABE, 2018) 등을 출판했다.

A value that arises from the boundary between private and public

by Jungmin Nam

1

As I take a stroll in a city, I often come across plants growing in the gaps within the walls and pavements. The instance when I started to take notice of these relatively unremarkable plants was when I realized that these urban city plants are a kind of a universal phenomenon that can be found in all cities across nations and regions. I was able to encounter these plants growing in such gaps whenever I was walking in various cities, whether in Seoul or in Cambridge.

The plants that I coincidentally came across by lowering my eye-level to the ground did not appear as sporadic and random but as having a substantial ecosystem of its own. I could indirectly confirm my suspicions on whether the role of these plants have a greater significance than it is usually conceived through the data researched by Peter Del Tredici and his students at Harvard Graduate School of Design. According to the research, about 9.5% of the urban area of Somerville, Massachusetts is occupied by these wild urban plants, and this is bigger than the area occupied by parks run by people.

As significant as the size of area that these wild urban plants occupy in a city, they also contribute to the city in many ways. The plants growing on pavements reduce the heat of the city, retain soil and water when it rains, removes pollutants, and acts as a habitat for micro-organisms. These plants that grow spontaneously in an urban environment have their distinct patterns when constituting their ecosystem. This invites considerations for methods that view the urban plants not as something to be removed in the construction process but as something to be embraced as part of the architecture itself.

2

Many describe Seoul as a 'republic of apartments', but going back to the 1970s and 80s, the main form of residence in Seoul was a detached house. Residential urban structures that did not end up as an apartment rapidly changed their forms from detached houses to semi-detached, multiplex residences, and these kind of residences still represent more than 50% of total residential types in Seoul. In other words, it may be said that at least 50% of Seoul residents live their daily lives in a densely-populated area with such a residential type where even the minimum landscape area requirement is not applied.

In such a physical environment that does not have any park, field, or a minimum landscape space – a place where buildings are filled up to the brim of the site boundaries out of personal financial interests – people autonomously find their gap spaces to allow nature to come in as well. These people locate gaps within their walls and site boundaries – some utilize the form and the width of the roads along their walls to locate a gap between the road and the traffic flow line – to put their vases and garden beds to grow their plants. This spontaneous landscaping that arises from the residents' desires without any intervention from an architect develops its own pattern through an engagement with the physical context and the invisible dynamic relationship between private and community. Through this, a virtual depth created by the plants is added to the expressionless, 2-dimensional walls and building facades, and a conversation between the road and the building begins to take place. These plants that were grown solely for personal interests become in turn a shared landscape and mini-gardens for the people in that region. Generally, in architecture, the surface is placed on the acute point of contact between the private and communal, between the building and the city and when the surface obtains a virtual depth from this, the gap space thus formed finds itself capable of embodying a new kind of possibility in an urban setting.

3

Historically, architecture has developed in an oppositional relationship to nature by trying to

overcome it. Buildings were built, cities were formed, and structures took over the place where nature originally had been. However, in the midst of this competition between buildings and nature, people have sought various ways to keep nature by their side in their daily lives. To escape from urbanization, some go look for untouched nature such as mountains and some have brought in nature to their lives by building parks in their cities. Some have secured a place for nature to be part of their lives by building personal gardens or flower beds, and if that was not feasible, they buy vases and pots.

However, these series of efforts to bring nature into their lives all happened beyond the realm of architecture. Is it because architecture has always focused on creating space and place for human beings instead of a place for nature? W hen planning a building, nature was either used to fill up the empty spots in the site that were distinguished from the buildings, or treated as a secondary priority to be considered only after construction. In architecture, nature was always treated not as its necessary element but only passively approached as an ancillary element outside of architecture.

However, there are traditional cases around the world where the integration between nature and architecture was actively sought. Most representatively, the sod roofs in residences of Northern Europe allow one to view the relationship between nature and architecture in a more unified manner. As a necessary element that composes the roof together with the regional climatic property and the wood-based construction method, the sod roofs of Northern Europe was one of the most widely-practiced indigenous architectural methods until the 19th century. It is a case where nature was approached as a core component of architectural composition and applied holistically with the building's exterior form, function, and construction.

4

The Mining Engine House, which had contributed towards the industrial revolution by leading the mining enterprise business in Cornwall, England around the 18th century, is left abandoned as an

empty shell since the decline of the mining business after the 19th century. These engine houses – which are now designated as UNESCO World Heritage Sites – are still as large and abundant in size and number as it was during the booming years, and it is still possible to spot these engine houses erected here and there in a sporadic manner as one strolls around in Cornwall. From the start, the Cornwall engine houses are buildings that have no relationship with a certain sociocultural agenda or respect towards nature. Rather, as a building that was fully intended to exploit natural resources and contribute towards industrial development, this structure was designed with a sole interest towards efficiency, such as on whether the size of engine and its location are appropriate for mining. Now, however, having stayed out for more than 100 years, these houses stand there in nature as a part of it, resembling a natural landmark created by earth during its geographical formation despite their intrinsic oppositional relation with nature.

Having been built with stones from the region, Cornwall's Mining Engine House is a structure that is equipped with the bare minimum functions to fulfill its role as an engine house. Because of this, it takes on a simplistic, rationally designed form that incorporates only the minimally necessary elements such as a chimney and a singular mass. This reveals the contrast between nature and an artificial structure. This creates a special kind of ambience that is difficult to be emulated by nature alone. It is necessary that the city embrace both buildings and nature. The Cornish engine house makes one rethink about the possibility and the new relationship that can be fostered between architecture and nature.

5

From the Tanpopo House designed by Terunobu Fujimori during the 1990s in the outskirts of Tokyo to the world's first vertical forest apartment building Bosco Verticale designed by Stefano Boeri during the 2000s, attempts to combine architecture with nature continue in the contemporary times. While the attempt to unify nature and architecture was achieved with the roof of the traditional buildings in Northern Europe, the

more recent projects since the late 20th century expand the subject of unification to the vertical walls of the building's facade.

Using vertical greening as a project feature, various vertical greening products that integrate modular potted plants and watering system are available in the market nowadays. Locally, these product lines are also being utilized in wall vertical greening projects for various cases such as for governmental buildings and shopping district interiors. However, in most of these wall greening systems, there has not been enough contemplation on the holistic relationship between vegetation and architecture; rather, instead of approaching the project with a focus on integration, the interest remains merely on the sole function of filling up the 2-dimensional vertical surface with plants, thus leaving the plants and the building disjoined as two mutually independent entities.

In contrast, Terunobu Fujimori's holistic attitude towards architecture and vegetation reveals his attempt to reconstitute the traditionally oppositional relationship between nature and architecture into an unified one. His architecture perceives nature's existence – which is represented by the plants that have been reduced as a mere part of landscape in conventional architecture – as a part of architecture and even expands the interest of architecture to include nature. This is still an ongoing experimental phase, but it makes one think of the possibilities that architecture may gain in the future through such an approach.

6

The shock that came to me when I first saw a zoomed-in photograph of the Guaranty Building

designed by Louis Sullivan was because of its plentiful ornaments. Because of my preconception of this building as the most representative manifestation of his extremely famous concept 'form follows function', the fact that there were beautiful and abundant ornaments in every corner of the building was a complete reversal of my expectations.

This early high-rise building – which is divided into four sections according to their respective functions and reflects the construction logics of the metal frame directly onto its exterior facade – displays a building composition that purely follows a principle of function. However, not unlike a classical building, it is fully ornamented and upon a closer look, it also features a depth in its surface by touting a new ornamental pattern that is distinct from the ones of previous generations. One may misread the quote 'form follows function' as to represent a dry, uniform, international style building, but the function proposed by Louis Sullivan was something deeper and inclusive of the sociocultural functions. Due to the terracotta decorations that cover the entire building – which may be misunderstood as something that is opposed to function – this building, while being different from the previous traditional architectures, puts on a friendly countenance and blends into the urban scenery. This new and unfamiliar foreigner with its high-rise element enters the city and overshadows its neighbors, but wears an amicable appearance and greets the surrounding roads and buildings. Thanks to these friendly compellations, it must have been possible for the neighbors (buildings) to shake off the uneasy feeling that comes from the relative scale difference and welcome it as a part of the city.

7

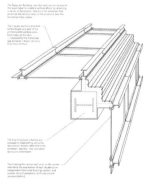

The vertical mullions of the Seagram building that is calmly clad in black look austere at first sight. However, as it is revealed from its detail drawing, this is not so much of a structural feature but an ornamental effect. We were able to reconsider the relationship between ornaments and a building's envelope through this.

The further interpretations regarding various ornaments of the building's envelope show that the ornaments do not only exist as ornaments but also constitute the structure system and are intimately related to the building's behavior method as the ornaments affect how the buildings respond to the changes in the external environment. The building's envelope that becomes conveyed through the ornaments acts as a medium between the building's insides and outsides, provides direct and indirect experiences of the interior space in regards to the exterior space, and influences the relationship between the inside and the outside.

Farshid Moussavi's 'Function of Ornament' does not limit the meaning of function to a function in the engineering sense such as a structure or a program; rather, by expanding it to include sociocultural functions, and the communication between the exterior and the interior of architecture, it provided the means for me to widen my view towards the building's envelope which I had merely conceived as a kind of a thin shell. The expression 'depth of surface' may not be directly mentioned in this book, but the countless successful buildings introduced here effect a different kind of function as architecture – that is, by possessing this depth of surface, these buildings fulfill their inevitable roles that they have to play towards their new neighbors as they enter into their contexts.

8

Learning from Las Vegas influenced me significantly in my view towards everyday architecture and the surroundings. Through this book, I was able to reperceive the fact that the everyday which I may simply take for granted can also be an object of study and discovery as well as a source of insight. It also led me to rediscover how much I can learn from things of everyday life. With Las Vegas as a study model, the authors in this book rediscover the role and value of architecture through an observation and analysis on general and commercial architecture. The authors propose an alternate view against the prevailing idea of modernism of that time. By reevaluating the predominantly-established value in the architectural world through everyday architecture, the authors' attempts to propose a new alternative by analyzing and organizing the value of everyday architecture came to me as quite meaningful.

Through this book, I pose a question to myself: is this process of rediscovering and looking for alternatives in terms of architectural value through the things learnt at Las Vegas in the 1960s and 70s happening again in the contemporary Korean architectural scene? Or has it ever happened before? I see this approach and attitude of the authors in this book as something universal across time and space, and as something essential to Korea of today. The process of observing and analyzing architecture and its attitude still works as my personal guideline as I deal with architecture.

9

How many buildings have we experienced by actually using them? How many buildings among the countless buildings in a city will we ever get to use and experience? If we were to narrow down our experience of buildings to this condition of physical encounter, most of the experiences would be simply about the surrounding building surfaces.

The process of experiencing the relevant media and contents as we read a book, admire a picture, listen to a music, or watch a movie can be emulated in a rather similar way to others. In contrast, the experience of a building can be further divided into the experience of actually using it and the experience of its surface. While the various experiences of everyday that permeate the building throughout are limited to a relatively small minority of its users, however, the majority of the population that shares the city or the neighborhood where that building is located at only gets to experience the building through its surface. And this experience is a one-way experience that is not based on choice.

A building may be a personal property, and it may be designed out of a personal desire, but through its surface, it eventually has influence over the life experience of the public. The communicative function that a building performs with its external elements such as the public and the urban city is conducted through its surface. To provide a more comfortable and pleasant physical environment and to have more inviting roads and places in a city, the surface plays a significant role that is as important as the interior space of the building. It may even be said that the role of the surface in such a dense urban environment with a high floor area ratio like Seoul is even greater.

The surface of the building plays an essential role in its relationship to the public. The building's surface is always in between the boundaries of private and public, and our city will become equipped with a better environment if we were to tap into its various potentials.

PROJECT

Living-Project

Everyday Architecturalization: from the vase to street furniture

The surface of architecture has always existed in our daily lives as the boundary where personal interests and urban communality strive against one another within the realms of building and city. Also, we experience architecture through the urban space created by these building surfaces (for e.g., road, plaza, etc.). Ironically, the public worth that the building surface embraces cannot but be taken more seriously in a densely-populated city such as Seoul.

A building may be built out of personal interests, but through its surface, it engages in a conversation with the city. However, when the surface goes beyond its 2-dimensional surface by adopting a sense of depth, more conversations can occur. In the neighborhood streets of Seoul, we often can find potted plants on the walls and wild plants growing in the gap spaces within the walls. Through this addition of plants on an otherwise plain wall, this wall that once divided the private and the public turns into a community experience for the people walking by that street. Putting its basis on these series of observation and experience, Living-Project is a project that searches for possibilities to approach these things in an architectural way. Through an architectural design method of approach that is simple and modular, the daily landscape was sought to be architecturally translated. This aim of the project was realized over various scales ranging from the scale of a potted plant or a piece of brick to the scale of a street furniture system.

Living Brick

A brick with a pocket at the sides for plant growth was proposed. The protruding gap space provides a space for plant growth on the brick surface, and allows for an ordinary construction activity and an ecological environment to be seamlessly integrated with the building's outer layer. An universal applicability of this construction process was secured by making it also compatible with regular bricks.

Architect: Jungmin Nam
Design team: Byung-gyu Joo
Use: building component
Size: 230×90×57mm(L×W×H)
Material: glass fiber reinforced concrete
Production: KSS
Design period: 2016. 11 – 2017. 1
Photograph: Gihun Noh

Living Block

Living Block is a project that utilizes the complex and various gap spaces of a single modular unit. A complex pattern and aggregation was made possible by combining a singular unit with various sides. Each side of the Living Block portrays an unique frontality, and while it can function as a polyhedral pot by its own, it also acts simultaneously as a building block that contains depth for vegetation.

Architect: Jungmin Nam
Design team: Byungsoo Kim
Use: building component
Size: 200×200×200 mm (L×W×H)
Material: glass fiber reinforced concrete
Patent: Korea 10-1701001 (Jan. 23, 2017), US 9,901,036

B2 (Feb. 27, 2018)
Production: prototype model
Design period: 2014. 4 – 2015. 1
Photograph: Gihun Noh

Living Pocket

Living Pocket provides a space for plants to grow by creating a gap in the thin facade. Through this, the original function of the building envelope extends to the realm of vegetation. Living Pocket is designed as panels so that it can be compatible with a typical stone wall construction method.

Architect: Jungmin Nam
Design team: Byungsoo Kim
Use: building component
Size: 600×600×150mm (L×W×D)
Material: glass fiber reinforced concrete
Production: KSS
Design period: 2013. 11 – 2014. 7
Photograph: Gihun Noh

Living Hole

Living Hole is a project that applies the unit combination and depth to a horizontal surface. As a basis, three units that differ from one another by their holes and shapes are formed, and a porous layer is composed horizontally through their combination. Like the gap between pavement blocks in the city where wild plants grow spontaneously, these three unit types made of concrete possess small grooves of different forms and sizes and act as a medium to contain nature at a small scale.

Architect: Jungmin Nam
Design team: Byung-gyu Joo
Location: Soda Museum, 30, Hyohaeng-ro 707beon-gil, Hwaseong-si, Gyeonggi-do, Korea
Use: building component
Size: 260×112×110mm(L×W×H)
Material: glass fiber reinforced concrete
Production: KSS
Design period: 2018. 2 – 4
Photograph: Gihun Noh

Living Puzzle

Living Puzzle is an example that expands the module-based architectural approach on the vegetation block onto the scale of street furniture. The module is formed by these three types: a form of an equilateral triangle, a parallelogram, and a trapezium. Through the calculated changes in the sizes and heights, each module is designed to combine and connect with one another to allow reconfigurations of these three units into various forms of street furniture. Functions such as vegetation, bicycle stand, and resting area can be adopted through these various height differences as they are also simultaneously adopted as a plant environment by the riverside.

Architect: Jungmin Nam
Design team: Joohee Jo, Deahan Kang, Hoe-eun Hwang, Youngmin Lee, Joohyun Yong, Jung-a Hong, Sungho Goong
Location: under the Yangpyeong Bridge, 502-1, Yangpyeong-dong, Yeongdeungpo-gu, Seoul, Korea
Program: street furniture
Size: 2700×1170×1150mm (L×W×H),
2338×1170×850mm, 1350×1170×850mm
Structure: steel pipe, glass fiber reinforced plastic
Exterior finishing: GFRP with paint
Construction: KSS
Design period: 2015. 6
Construction period: 2015. 6 – 7
Client: Seoul City Government
Photograph: PACE Studio

Alley House

As a project for multi-family houses and neighborhood living facilities, Alley House is situated at an area in Banpo-dong where it is densely packed with multiplex and neighborhood facilities. Nature was always pushed back in terms of priority in Seoul's rapid urbanization process, and this marginalization of nature appears most manifestly in such dense neighborhoods where different kinds of interests intermingle. Due to its scale of size where the minimal legal landscape area does not apply, the bare minimal legal distance between buildings is kept in this area in an urban structure composition.
If one looks around carefully while taking a stroll in this town, one can find green life growing in the streets, gaps and corners between buildings, or even in isolated areas where the hands of the architect or the public have not reached.
The spontaneous greening and the landscape-patterns that came out of the residents' desire for greeneries manifest themselves in these isolated gaps with a kind of an orderly structure of their own.
With the observation of this spontaneous greening that appears autonomously in the streets as its base of interest, the Alley House project provides a gap for rest and nature in the building's surroundings and public areas. Reflecting on the densely-packed neighborhood living conditions, this project overcomes the local limits by offering an alternative to come in daily contact with nature – albeit at a rather elemental level – in the neighborhood facilities.
The facade that faces the alley as a main entrance that bends and enters from a 8m street functions as a new face for the building. The entrance facade, which is composed of red and blue old bricks and Living Bricks, forms a slight angle with the 8m street and thus creates a depth and a gap which acts as a space for green life to grow. The depth of surface and these green spaces facing the street function not only as the facade of a private site but also a public landscape as they become simultaneously reconstituted as a public environment that infuses a breath of life into the streets.
The main entrance alley is linked vertically into the building through an open external staircase that is formed in the gap between the studio rooms.
This staircase functions not only as a means of movement, but as a small park for people who have to live in studio-room conditions. In this gap space created by adjusting the width of the staircase and the stair landing, a public space where plants can grow and people can rest becomes created.
This method of approach that bases itself on the spontaneous growth patterns without architectural intervention proposes a possible alternative even in an adverse condition such as of a staircase.
Through this, the alley that was once dark and inaccessible becomes a welcoming street and an improvement to the urban organization.

Architect: Jungmin Nam
Design team: Hongryang Lim, Byung-gyu Joo
Location: 725-13, Banpo-dong, Seocho-gu, Seoul, Korea
Program: housing, commercial
Site area: 135.9m²
Building area: 81.22m²
Gross floor area: 260.01m²
Building scope: 4F
Height: 14.06m
Parking capacity: 4
Building coverage: 59.76%
Floor area ratio: 191.32%
Structure: reinforced concrete
Exterior finishing: brick, GFRC block
Interior finishing: wood, gypsum board with wallpaper, tile
Structural engineer: Daereung Architects
Mechanical engineer: Daeo ENG
Electrical engineer: Daeo ENG
Design period: 2016. 7 – 2017. 1
Construction period: 2016. 10 – 2017. 7
Client: Kyutae Chung
Photograph: Kyungsub Shin

Yellow Foot

Yellow Foot is a solitary apartment project located in Seocho-dong where it is densely packed with multiplex residences and neighborhood facilities. It is difficult to find buildings engaging with the streets in this neighborhood. Because the buildings block their boundaries with walls or have indifferent attitude towards the streets, these buildings are disconnected from the streets like islands. The neighborhood also lacks public facilities such as parks. A small playground functions as the sole public space for this area, but this is insufficient to

fully cater to the neighborhood. In this situation, Yellow Foot apartment was proposed as an attempt to encourage a more outreaching and engaging relationship between the buildings and streets.

Following the local practice for efficiency, on the ground level, the piloti was introduced to secure parking space and the floor area ratio. The piloti space functions as a main entrance interacting with the street by extending the surface area of the ground entrance toward the streets with the surrounding landscape. Through this interlocking approach between ground level and street, an entrance for automobiles and pedestrians was designed to provide an apartment's threshold that is open towards the neighborhood in a communicative way. Unlike other neighboring buildings, by having the surrounding landscape permeating into the street, Yellow Foot creates an open boundary with nature which brings about vitality to the dreary local streetscape.

The building facade has two different types of windows, one for the living room and the other for the regular rooms. The windows that are following shifting patterns in accordance with a granite panel size superimpose arbitrary patterns over the orderly facade. The inner space is also divided into flexible and rigid areas along the outer walls to accommodate the shifting window patterns with efficiency so that the outside openings and inside space are coordinated in harmony. The granite for the exterior cladding has a chiseled finish and grooving patterns which give an additional depth to the surface. On the stone surface, the three grooving patterns and their combinations brings about a rich and complex appearance of the building towards the street, by adding its shading patterns and color coded window frames.

Following the setback regulation that abides by the sun exposure plane and the local regulations of Seocho-gu district, the building is formed by

vertically stacking masses of two distinct sizes. Despite the restrictive conditions of the development-oriented project often found in cases of solitary apartment development, Yellow Foot apartment devised a way to diversify unit composition. Through the core that is shifted to the side from its center of the building mass and the entrance locations for automobiles and pedestrians, six different units based on 2-bed and 3-bed units were created.

Yellow Foot apartment provides an open boundary amongst the neighboring buildings that remain disconnected from the streets. Through this urban gesture, it proposes an alternative to the relationship between streets and buildings in both physical and visual approaches. With the depth of the elevation facing the street as well as the outreaching boundary with landscape, Yellow Foot apartment will encourage favorable relationships between the private and public area in this development-oriented neighborhood.

Architect: Jungmin Nam
Design team: Hongryang Lim, Byung-gyu Joo
Location: 1344-3–6, Seocho-dong, Seocho-gu, Seoul, Korea
Program: apartment housing
Site area: 790.9m²
Building area: 343.35m²
Gross floor area: 2,591.63m²
Building scope: 10F, B1
Height: 32.2m
Parking capacity: 31
Building coverage: 43.41%
Floor area ratio: 249.51%
Structure: reinforced concrete
Exterior finishing: granite, stainless steel, paint
Interior finishing: wood, gypsum board with paint, tile
Structural engineer: Quantum engineering
Mechanical engineer: Chunil ENG
Electrical engineer: Chunil ENG
Construction: Yemi Construction
Design period: 2015. 7 – 2016. 7
Construction period: 2016. 1 – 2017. 4
Client: Prestige
Photograph: Kyungsub Shin

Flower+Kindergarten

Flower+Kindergarten began with conditions similar to other urban kindergartens that strive for the maximum class room size and number of students. The core of this project was to provide an architectural solution that can fulfill those requirements while escaping from the standardized double-loaded corridor plans. It was a situation that called for a balance between the

value of the floor area ratio and the value of an educational space. While securing the maximum volume we propose a kindergarten where the children can not only find stimulation in their creativity from the spatial abundance but also experience nature in a more intimate way.

The solution was to create depth in the surface. By giving depth in the building's outer layer, gap space for plants and depth in the windows were created, and by forming an internal promenade that is moving along with the spiral openings, the relationship between building interior and the exterior was emphasized. As a spiral staircase, the internal promenade is connected to the hall at different locations per level and thereby provides a 3-dimensional connection to all floors. This creates a massive playground formed out of slides, staircases, and gap spaces. From their daily activities involving such vertical movements, the children come to experience spatial abundance, visual stimulation, and a 360-degrees panoramic view of the surrounding sceneries throughout the seasons.

By placing three classrooms and a multipurpose hall at each floor, a kindergarten without corridors was built. The multipurpose hall embraces various functions and movements as a second classroom and a playing space. The classrooms surrounding the hall are defined by non-load bearing walls, and the inner wall creates a 3-dimensional form of a classroom by adopting curved lines. Through this, the classroom takes on a distinct form, and through its inner wall of a certain thickness, a multipurpose gap space where it is possible to store things and play with is created.

The interior is classified by different colors at each floor, such that one can perceive and differentiate space through colors such as yellow and pink. An environment where one is led to pick up the subtle difference between colors is established by applying different shades of the same color group according to the architectural element. The color of the interior flows out towards the exterior of

the white building through openings and thereby creates a new expression and image on the building facade. Through the mass that abides by the site boundary, the continuous ribbon window belt, the large and small windows that are meticulously positioned according to the scales of a child and an adult, and the color that seeps out through them, Flower+Kindergarten has become a small but new landmark that functions as a guide to a repetitive apartment district.

Architect: Jungmin Nam
Design team: Byungsoo Kim, Jungsoo Seo
Location: 731, Umyeon-dong, Seocho-gu, Seoul, Korea
Program: education facility
Site area: 608m²
Building area: 303.44m²
Gross floor area: 2165.36m²
Building scope: 4F, B2
Height: 18.5m
Parking capacity: 8
Building coverage: 49.91%
Floor area ratio: 199.61%
Structure: reinforced concrete
Exterior finishing: crema bella, GFRC panel
Interior finishing: wood, gypsum board with paint
Structural engineer: The Kujo
Mechanical engineer: BOW ENG
Electrical engineer: BOW ENG
Construction: Yemi Construction Company
Design period: 2013. 5 – 2015. 1
Construction period: 2013. 10 – 2015. 1
Client: Yewon Kindergarten
Photograph: Kyungsub Shin

CRITIQUE

The Depth of the Surface and the Depth of the Root: the Young Architect Jungmin Nam

by Hyon-Sob Kim (professor, Korea University)

"Surface is the most accessible and shared element in a building, and it is in touch with our daily lives across the various realms outside and inside of a building. The buildings outer surface has always existed as part of our daily lives, functioning as the spearhead party at the borderline between private and public realm in that subtle point where the building and the city come in contact. In most cases, we only experience the surface of a building. Because we come to experience the urban space through the public space created out of multiple surfaces of distinct buildings (street, plaza, and others),

paradoxically, the architectural public value that a surface possesses in a development-driven, high-density city such as Seoul can become more significant. Although a building is built from a private interest, it engages with the public through its surface, and when the surface goes beyond its 2-dimensionality to hold a sense of depth, a deeper conversation becomes possible. The aim is to expand the boundary of surface that lies at the intersection between private and public ownership to the realms of the building facade, interior, flooring, and streetside, and to investigate the architectural possibilities within those minute gap spaces which are generated from contemplations on depths that may also be applicable to the surface."
— Jungmin Nam, 'Depth of Surface', 2018. [1]

What if architecture is about the surface?

Jungmin Nam's decision to promote his portfolio under the theme 'Depth of Surface' may be read as an appropriate reflection of the recent developments in architecture. For a while, in contrast to other core architectural elements such as space or structure, surface has always been pushed aside as something secondary and insignificant. However, the reality is that even cases such as Le Corbusier's 'Facade libre' or Robert Venturi's 'Decorated Shed' are too an expression of interest towards the surface.
This is also evident from our more proximate observations. There are already numerous contemporary architectural issues that deal with the surface as their interest — whether they are related to a phenomenological interpretation discussing how our bodily sensations connect with the building's outer layer matiere; or to a kind of play derived from media facade made possible through digital technology. This interest in the surface may be said to have continued from the discussions that David Leatherbarrow and Mohsen Mostafavi had introduced decades ago through their book *Surface Architecture* (2002). While focusing on the irreconcilability between production and representation expressed onto the building surface, these two individuals sought for possibilities to practice and introduce a new form of architecture. The 'surface effect' shown by Herzog & de Meuron with their Ricola Storage Building (1987) and Eberswalde Library (1996) is a representative example of such a possibility. [2] In this way, I think that it would be possible to find a sufficient 'depth' in the architectural surface if we were to adjust our perspectives by just a little.
This is what came first to my mind as I reflected on what the young architect Nam had meant by 'Depth of Surface'. Considering especially the fact that the dean of Graduate School of Design at Harvard where Nam had studied at was Mostafavi, it is natural to attribute 'Surface Architecture' as a part of his theme. However, when asked, Nam told me that

he only got to know of this book after he already had interest towards surface, although he did admit that the word 'surface' and his confidence on the topic was indeed derived from the book. [3] He indicated that it was the books by Farshid Moussavi and Antoine Picon — classes of which he took at Harvard GSD — that inspired him to take interest in the surface. Moussavi, who is a founder of Foreign Office Architects (FOA), analyzed the various types of ornaments as the medium for specific effects in her *The Function of Ornament* (2006) [4]; and Picon, being encouraged by Moussavi, introduced new potential meanings by 'the return of ornaments' in his *Ornament: The Politics of Architecture and Subjectivity* (2013) [5]. (Ornaments that have been removed since Adolf Loos in the early 20th century has returned!) In fact, Nam said that he only partially referred to those two books. However, he had interest in contents related to digital design and fabrication. It is assumed that this interest was renewed by the ornament or the surface effects that arose from such technologies. Picon's book begins with this sentence: "What if architecture is ultimately about ornaments?" Seeing how he quotes a poet writing that "nothing is in truth deeper than the surface" a few paragraphs down, this sentence may be changed in this manner: "What if architecture is ultimately about the surface?"

Practice of Surface

On whether architecture is ultimately about the surface is something to be studied further; however, in architecture, things like space, composition, and tectonics still remain as the more prioritized core issues. Nonetheless, it is undeniable that the importance of surface is being reassessed in the current architecture world. Nam is an architect residing precisely in that same flow although he appears to desire for an even more ambitious theme. [6] If so, how is his current design 'practicing' the surface?
Having returned to Korea in summer 2013 after his studies and work in the US, he has been building his career and a robust amount of work experience. While managing an office with a team of individuals, he began working on September of that same year as a professor at Seoul National University of Science and Technology, opened his own architecture laboratory OA-Lab (Operative Architecture Laboratory) soon after, and is currently active as a public architect of Seoul since 2015. However, it is obvious that what is important is not such official titles, but the kind of designs that he has created thus far. Over the five years since his return, he has realized three blocks of middle-sized buildings: that is, Flower+ Kindergarten (2013–2015), Seocho-dong apartment (2015–2017), and Small Garden (2016–

324

2017). His debut project, Flower+Kindergarten, which is located at Umyeon-dong, Seocho-gu, is impressive for its flowers on the wall surface and it has not only been published in *SPACE* (Feb. 2016) and elsewhere but has also received awards.[7] Seocho-dong Apartment is a single 9-storey apartment block that holds 28 households, and it has been termed in English as 'Yellow Foot' for its yellow pillar at the piloti space on the ground floor. On the other hand, the identity of Small Garden at Banpo-dong, which is actually a medium-scale multiplex residential building, is more correctly perceived through its English name, the 'Alley House'. The first floor is planned to hold a store, the second and third floors are planned to hold six units of studio apartments in total, and the fourth floor is planned as the residence for the building owner. In addition to these three buildings, Nam also showcases what he calls as Living Project, which is a set of plant pots and street furniture made out of various bricks and blocks. Amongst them, his Living Block, which was conceived in 2015, is a unique potted plant block unit that has recently obtained patents from Korea and the US, and these blocks can be put together to form a wall. These are the works that have been submitted to this year's Young Architect Award Design Competition, and his partial remodeling project of Songcheon-dong Community Service Center that he did as a public architect and his other public architecture designs have not been included among them. For a young architect, this is quite an impressive list.

As seen from the aforementioned buildings, what is immediately visible as a common feature of his architecture is his use of bright colors. Without doubt, yellow is his base color palette. When the Flower+Kindergarten was finished, I assumed that his reason for using various pastel tones for each floor was because it was a kindergarten. However, not only in his Yellow Foot and Alley House, but also in his remodeling work of Songcheon-dong Community Service Center, one can find bright yellow patches that indicate Nam's architectural identity. His color application does not seem to follow a strict code, but a code nonetheless. This is because the plane that cuts the vertical outer layer horizontally and the setback space towards the inside of the outer layer appears as the main highlight clothed in color. In the case of Yellow Foot, the former applies typically to the inner plane of the window frame, and the latter applies typically to the roof, wall, and pillar surface of the piloti space on the ground floor. While the emphasis was placed on the feature of using different colors at each floor in Flower+ Kindergarten, which shine out through the spiral staircase winding up the building and the window belt, the same logic applies to the large windows placed here and there and the decision to color the inner plane of the entrance frame. Also, this logic of Yellow Foot was also applied somewhat lightly in the case of Alley House as well, and it is especially worth to mention how the yellow color is expanded from the first-floor piloti space to the staircase room that connects the four floors. Whether it is the piloti space for parking, or the staircase room, let us remember that these spaces are all public spaces within the building. To describe Nam's use of colors generally and metaphorically, he tends to add colors what is revealed inside by making a cut in the outer skin. To reiterate, in Nam's architecture, the colored parts are the places where one can most intuitively sense the 'depth of surface'.

Let us expand further on these two subthemes – that is, the window frame and the public space – that we have gathered from this insight of the building and its inner color. First, the window frame. As seen in his Flower+Kindergarten and Yellow Foot, before coloring the insides of the windows that he chose to emphasize, the frames were made to protrude first. Through this, the window frame (including both the lintel and the sill) gains sufficient depth and leads one immediately to the theme on depth of surface. The protruding window frame also fulfills a practical role as a kind of a louver that covers the sunlight, and in the case of Yellow Foot, this will also be accompanied with a function to block out unwanted views from the neighboring houses. Perhaps an unconscious move out of the desire to be compensated for the missing balcony? Regardless, what is more meaningful from my perspective, however, is that a rich expression had been added to an otherwise flat facade by adding depth to the building surface. The brightened expression of the building surely finds greater significance not only in terms of its own improvement but especially in terms of how it connects with the surrounding urban context. The protruding frame of Yellow Foot, which is made by an elegantly thin metal plate, creates a contrast with the deepened surface and brings about a refreshing sensation, although it would be difficult to compare it with the radical experimental nature of MVRDV's Wozoco Apartments(1994–97) in Amsterdam. Nonetheless, this method of giving depth to the surface by either protruding or retreating the window frame or plane is interesting in that it stands on the opposite side to the alternative experimental method of erasing all sense of depth in the building's outer layer. In the Korean context, this reminds one of Seung H-Sang's numerous designs, such as his Welcomm City (1995-2000). In that project, by perfectly aligning the building mass's Corten steel outer layer and the frameless window surface to the same level, Seung created an effect that cancelled out the outer wall's weighty appearance.

The Publicity of the Surface and the Surficial Public

On the other hand, the yellow-theme public spaces reveal the building's insides in a deeper way than the window frame. As mentioned above, the first-floor piloti space of Yellow Foot and the staircase room of Alley House are such examples. It may seem a little unnatural to expand this discussion on building surface to a space so deep inside the building, but surely, space is created by putting a surface together with another surface. The depth in the window frames too already incorporate a spatiality. The paragraph that I cited at the beginning from Nam speaks of a 'public space' constructed by a composition of surfaces. Also, this eventually arrives at the issue of publicity: that is, the public value of the crossroads between architecture and the urban city. Nam's hesitation to stay merely within the surface as the theme perhaps out of a desire for something more ambitious must have brought this about. Moreover, let us not forget that he is a public architect. First, there is a need to reflect on the two layers of meanings that the word 'public use' indicates. The obvious meaning is to share something and not keep it for private use only. Because both Yellow Foot and Alley House are shared houses, it is not unusual that they have public spaces such as carparks, stairs, and corridors. Building residents share these spaces, and it is hoped that a sense of community would develop from there. This is the first layer of 'public use'. However, can it be open to a real public including an indeterminate number of strangers? Perhaps the building may be open to the building management staff, delivery workers, or specific individuals who are related to the residents in some way, but it may be a completely different matter to complete strangers within this society of capitalism and private ownership. In this sense, the word 'Garden' in 'Small Garden' (a Korean name of the project Alley House) must have meant to indicate a 'a garden for community' and not a 'garden for everybody'. Only with this understanding of the limits within reality, we can then proceed to read the 'publicity' of the surface in Nam's architecture. Such publicity is at times also superficial.

The first-floor piloti space of Yellow Foot is first and foremost visually public. Other than the pillar and a small part of the core, everything else is visually accessible from the front road to the flower bed at the back. The neighboring apartment blocks at the left and right also have their piloti spaces, but because they form boundaries with a low wall or a screen, they do not possess this sense of freedom. Also, the bright

yellow color of the piloti space and the flower bed is bound to be eye-catching to the people passing by, and the park built in the flower bed in front of the building provides refreshment to the dreary urban space as both a 'garden for community' and a 'garden for everybody'. This bright and fresh aboveground appearance is further emphasized through the paradoxical lightness created by the building's surficial depth. What I mean here by 'paradoxical lightness' is to indicate the aforementioned description of the sensation that arises from the thin window frame, and the thin solid grooved line patterns on the outer granite surface also add towards this effect. After its construction, Yellow Foot brought about an interesting phenomenon from its surroundings. A yellow clothes collection box was placed in front of the building, and the wall of a multiplex residence across the street was painted in yellow – in a sense, yellow was being spread in the neighborhood. Regarding this, Nam observes it to be 'a result created from a conversation between the aboveground public space and the road'. [8] The staircase room of the Alley House is a vertical extension of the piloti public space. In this staircase room, the intention to bring in the horizontal system of the urban city into the building interior is very clearly present. To someone who is familiar with Korean contemporary architecture, this intention is not new. Furthermore, isn't what Nam refers to as 'the extension of the alley' [9] something that we have heard before quite often? For example, Sungyong Joh had described the staircase of Yangjae 287.3 (1990–92) as 'the extension of the road', and Seung H-Sang had expressed the yard of Sujoldang (1992–93) as 'the extension of the alley connected to the city road'. [10] We can also find instances where this architectural vocabulary was used in shared residences, and the most representative ones are Bang Chul-rin's multi-household residence STEP (1994–95) and E Il Hoon's multiplex residence Toegyebulyi (1996–96) and Gagabulyi (1995–96). Also, there is a need to bring up E's concept of 'house sharing'. [11] This concept that argues that it is necessary to create a communicable space between the neighbors by sharing a portion of the house no matter how small the house may be resonates with Nam's idea to create 'a niche shared space'. By halving the outer layer bricks, the increased surface area [12] of the building eventually provided for the private space that was lost from securing the shared space. On whether he was aware of the theories and the practices of these early architects of the 1990s is not so important here. This is because it already shows that the previous generations have carved the values of the public and communality deep within the Korean architectural society, and that the current generation is borne out from this foundation. The unique thing about Nam, however, is that he utilized the secured gap spaces in the staircase room for green vegetation.

Architecture and Plants

Indeed, the uniqueness of Nam's architecture lies in his efforts to integrate plants into architecture. By this, he distinguishes himself from not only the Korean architectural world, or from the young architects in Korea. While this distinguishing feature may not be as pronounced in Yellow Foot where the building and the flower bed is somewhat divided, the integration of plants is clearly evident especially when it comes to his Flower+Kindergarten and Alley House. While this is intimately related to the depth of surface, this can already be appreciated by itself as a distinct architectural method without even needing to mention the surface.

Let us look at his Flower+Kindergarten. As I remarked earlier that it is 'impressive for its flowers on the wall surface', the flowers that decorate the outer wall surface of the first floor and the series of pot panels that carry those flowers are the biggest features of this building. The name of the building was obviously coined from these features, and when compared with the gardens of the basement interior and the rooftop, the latter gardens look complementary. Nam called these pot panels that are attached on the walls as Living Pockets, and the aforementioned Living Project began from this. Living Pocket uses a concrete cubic unit composed of 60cm square panels per side with convex centers by using a smooth curved surface, and a pocket space is prepared inside to collect the soil and the plant while allowing for drainage by having holes below the pocket as like any other plant pots. Also, when multiple units are connected, it was designed so that the upper part of the pocket and the dividing paths of the water below the holes are well-aligned, and horizontal grooves were added at the center panel so that each unit can fit well together even when the panels are not aligned. Such Living Pocket units are connected and attached to the kindergarten wall, and they create a pattern that uniquely resembles either the waves or folds. Also, due to the pattern's 3-dimensional surface [13], the depth of the surface that can accommodate to hold flowers and plants goes beyond mere material depth to give a sense of emotional depth. Moreover, the flower planting activities on these well-designed walls for emotional cultivation and creative education of children would be something that is only possible in this kindergarten.

By applying Living Brick — one of the items in Living Project — in Alley House, a special feature is added into this building. Living Brick is made by filling only a half of the brick to let the rest of the brick function as a plant pot. Nam inserted the Living Bricks with a focus on the left side of the first floor and positioned them somewhat sporadically on the lower half of the second floor as well. By doing this, he opened up a possibility for plants to grow on the building walls. Adding to this, he installed a small flower bed at the 'deep surface' created by retreating the outer wall of the front and the left side, and all these combine to literally create a 'small garden'. On the other hand, the greening of the aforementioned staircase room is also an eye-catching element. The gap between stairs and the stair landings allow for plant pots to be placed or hung, and the series of wires that penetrate rails of multiple floors allow for vine plants to climb up the stairs. Also, although not as special, even in between the pavement blocks of the leftover plots at the front and at the carparks at the sides and back were designed to allow wild plants to grow naturally. No specially-designed block was used here, but a similar concept is later developed into a flooring block named Living Hole. This has been submitted as an entry piece for the currently ongoing exhibition 'Artificial Nature: Embodying Nature in Concrete' (Apr – Oct. 2018) held at SoDa Museum in Hwaseong, Gyeonggi-do, and it features a pavement block with holes for plants to grow in. Among the things that Nam had designed in his Living Project for purposes outside of his buildings, Living Puzzle is the most representative. This street furniture that was installed in summer 2015 under Yangpyeong bridge in Yeongdeungpo-gu is a curved bench with parts that can be put together in various ways, and it features an integration with plants by having not only its own bicycle stands but also flower pots. While the idea of integrating plants into architecture may have been Nam's main interest since his graduate project at Harvard GSD titled under 'Urban Farm, Urban Epicenter' (2009) [14], but this trend is not wholly new. Interests towards ecosystem recovery from environmental degradations have already been realized via various attempts in architecture and urban design. For example, the French botanist Patrick Blanc (1953–) who is famous for his vertical gardens have engaged in cooperative projects with various star architects such as Herzog & de Meuron and Jean Nouvel. Despite the apparent difficulties in terms of its sustainability, there were also various local attempts such as Minseok Cho's Ann Demeulemeester Shop (2007) in Sinsa-dong, which has attracted views for its elegant wall greening. However, for some reasons, Nam's design forms a stronger resemblance with the design by Terunobu Fujimori (1946–), who is a Japanese architect and architectural historian known for his very unique design as showcased from his floating tea house. From the beginning of his career as an architect, Fujimori was already integrating plants into his architecture and executing his principle of mass-utilizing simple technologies from the stone age. One is reminded of his 'Tanpopo House' (1995), which is a residence built with dandelions planted on the walls and the roof; his 'Nira House' (1997), where pots of leeks are inserted in a row on the roof; and his 'Tsubaki Castle' (2000), where the roof is entirely covered with grass and a camellia tree is planted at the peak. [15]

There are some reasons why I associate Fujimori with Nam. First, in a big picture, Fujimori's emphasis on the 'finish' rather than the space or structure is similar to Nam's commitment towards the surface. Second, in contrast to the wall greening as shown by Blanc, the fact that there are relatively less technological interventions in Fujimori's roof plants, and that his plants do not overtly overwhelm the architecture is quite similar to Nam's approach. Third, both Fujimori and Nam received their inspirations from a familiar objet that anyone can come across. This requires an extra explanation. Nam's portfolio begins from photographs of various back alley sceneries in Seoul that we all come across on a daily basis, and in those photographs, various potted plants and vines are forming a 'spontaneous vegetation'. Nam saw that this kind of vegetation adds a 3-dimensional surface to an otherwise normal building by the street, and he discovered the potential to do his own architecture from this everyday architecture that had no architectural intervention. It may not be the same, but his curious interest towards the everyday is quite comparable with Fujimori's effort to observe and record down everyday objects and sceneries by walking around the urban back alleys via his 'Architectural Detective' and 'Street Observation Society', although Fujimori was somewhat reluctant in directly relating these activities with his designs. As shown from this, however, there is quite a difference between these two individuals regarding plant-building integration as well. Although both individuals minimized technological interventions in the processes of planting and tending, however, Nam made full use of the digital technology to produce his plant pot units. Such cutting-edge technology was needed to produce the curved forms for Nam's Living Pocket units, and this goes completely against Fujimori's method which only utilizes stone age technologies. Such differences, however, are natural considering the different environments and educational backgrounds that these individuals were brought up in. Moreover, there is also a significant difference in what these plants within architecture

mean for the two architects. The meaning direction seems to go in reverse to how much each has utilized technology. This is because while the flowers and plants indicate a simple life to Nam, for Fujimori, however, these plants symbolize the ruins of technological civilization buried under the history of civilization. [16]

Concluding Remarks: Depth of the Root

In this manner, a reflection on Fujimori provides clues on how the plant-building integration of Nam's architecture may be further expanded in the horizon of meaning. The book *Plants & Architecture* (2008) edited by Taro Igarashi can also be helpful in this context. This book is a collection of various aspects on plants and architecture from a Japanese perspective, and it also contains a chapter on Fujimori [17]. Nam was not acquainted with or consciously aware of Fujimori, but after our discussion, he discovered a photograph of Fujimori's 'Nira House' among his reference materials for his graduate project at Harvard. In that sense, Fujimori was not completely irrelevant to Nam's conception of 'vegetation' during his studying years. Nevertheless, what must be remembered is that (just as in the case of Minseok Cho) whether in the case of Fujimori [17] or Nam, the integration of plants into buildings have not been completely successful. In other words, the integration process is still experimental. As how the dandelions and the leeks planted on Fujimori's roofs had shriveled, the flowers and plants in Nam's Living Pockets and Living Bricks also too lost their vigor and wilted away. However, seeing how the wild plants – regardless of the intentions of the two architects – are proliferating so well, and how the vine plants in Nam's Alley House are still surviving, it seems that the choice of plants is important. What is more important, however, and also very obvious, is that a sufficient amount of soil and depth should be provided for the plants to take root. Only then can the plants support themselves, thus making maintenance easier and the project sustainable. This point is ultimately connected to this text's overall theme. The 'depth of surface' should be added for the depth of the roots. Going further, contemplations on the surface should also take root deeply in the water streams of architectural history and discourse. Only then can such contemplations bear fruits regardless of whether they cover the disharmony between 'production' and 'presentation' or touch upon the gap between daily life and the base of civilization. Nam is still young, and many things are still expected from him. When the roots of his contemplation have grown deeper, I'd like to write another text on him, but this time as a 'blossoming architect': "A tree whose roots are deep: in the wind does not shake; its flowers have luminance; its fruit, fragrance …" (Translated by Keunho Hong)

1) Cited from his portfolio titled under 'Depth of Surface: building and the urban, on the generation of potential gap space between the private and public' which was submitted for application to 2018 Korean Young Architect Award.
2) David Leatherbarrow and Mohsen Mostafavi, *Surface Architecture* (Cambridge MA: MIT Press, 2002), pp. 209–214.
3) From personal exchanges of conversations and emails with Nam over 17–18 July, 2018.
4) Farshid Moussavi and Micael Kubo (ed.), *The Function of Ornament* (Barcelona: Actar, 2006).
5) Antoine Picon, *Ornament: The Politics of Architecture and Subjectivity* (Chichester: Wiley, 2013)
6) The theme 'surface' seems yet provisional for such a 'young' architect. From personal exchanges of conversations and emails with Nam over 17–18 July, 2018.
7) He was awarded with the Honor Award for Architecture at the AIA International Region Design Awards in 2015 and the First Prize at the Seocho Architecture Awards in 2017.
8) Jungmin Nam, 'Depth of Surface', 2018.
9) Jungmin Nam, 'Depth of Surface', 2018.
10) 4·3 Group, *Echoes of an Era* (Seoul: Ahn Graphics Publishers, 1992).
11) Regarding this point, see my article: Hyon-Sob Kim, 'The spectrum of the 4·3 Group Architecture and Critical Modernism', in Papers and *Concrete: Modern Architecture in Korea, 1987–1997*, edited by Dayoung Jeong and Sung Kyu Jung (Seoul: National Museum of Modern and Contemporary Art, 2017), pp. 78–87.
12) He vertically halved all the outer bricks for precise brickwork, and this brought about an effect where the width was widened by approximately 5cm in all sides of the building.
13) There are a number of similar preceding works, among which one may be reminded of Chan-Joong Kim's Han River Tunnel Wall Design (2009) and his Raemian Gallery (2009) in Korea. The way how he designed the units with digital instruments and pieced them together on site after having them mass produced at the factory is quite similar to Nam's Living Project.
14) He wanted to propose a high-rise vertical agricultural structure as an urban infrastructure and hoped that it would go beyond crop production to function as the epicenter for a sociocultural transformation.
15) I have written multiple articles on Terunobu Fujimori's architecture. In particular, see these two articles: Hyon-Sob Kim, 'Exciting Architectural Adventures of the Future Boy 'Terubo''. *SPACE* 536 (Jul. 2012), pp. 16–21. Hyon-Sob Kim, 'The uncanny side of the fairy tale: post-apocalyptic symbolism in Terunobu Fujimori's architecture', *The Journal of Architecture*, vol. 21, no. 1 (Feb. 2016), pp. 90–117.
16) Hyon-Sob Kim, op. cit.
17) 五十嵐太郎 編, 『建築と植物』(京都: INAX, 2008).

Hyon-Sob Kim studied modern architecture at the University of Sheffield, UK. Since his appointment as a professor at Korea University in 2008, he has been teaching architectural history·theory·criticism and now getting interested in writing a critical history of modern architecture in Korea. His recent publications include *Architecture Class: History of Western Modern Architecture* (2016), a Korean translation (2017) of *Building Ideas: An Introduction to Architectural Theory* by Jonathan Hale, and "DDP Controversy and the Dilemma of H-Sang Seung's 'Landscript'" (*JAABE*, 2018).

에필로그

2018 젊은건축가상
심사 총평

이은경 · 심사위원, EMA건축사사무소 대표

2018 젊은건축가상에는 1차 31개 팀이 지원했고, 2차로 7개 팀이 공개심사를 거쳐 최종 3개 팀 수상자가 결정되었다.

김이홍(홍익대학교 교수)은 다른 문화적 콘텍스트를 넘나들며 주어진 것에 대한 관찰을 개념화해 섬세하고 완성도 있는 결과물로 만들어냈다. 개념과 구축의 경계를 오가는 집요한 작업 과정이 돋보였다. 김효영(김효영건축사사무소 대표)은 건축의 역사성을 현재의 시간과 지역이라는 장소에서 재해석해 시각적으로 드러나는 질서와 기하학의 인상적인 작업을 보여주었다. 남정민(서울과학기술대학교 교수)은 공적·사적 경계면, 그 틈의 공간에 대한 지속적인 관찰을 공업화된 유닛의 실험을 통해 구축하고, 이를 작은 단위의 스케일에서 건축 공간까지 이어지는 다수의 프로젝트 변주로 작업의 일관성을 보여주며 앞으로의 발전을 기대하게 했다. 문주호, 임지환, 조성현(경계없는작업실)은 부동산 개발논리로 지어지는 보편적인 건물에 대응해 상황과 조건에 대한 논리적 분석을 통해 나온 설득력 있는 해결과 그 결과물의 완성도를 보여주었다. 특히 이를 건물을 짓기 위한 기본적 정보를 데이터 기반 자동화 프로그램을 통해 다수가 공유하는 방법론으로 제시한 점이 인상적이었다. 박혜선, 오승현(건축사사무소 서가 대표)의 주거에 집중된 포트폴리오는 보편적인 주택시장에서 역량 있는 젊은 건축가의 작업이 폭넓고 주도적인 역할을 할 수 있음을 보여주었다. 윤한진, 한승재, 한양규(푸하하하 건축사사무소 대표)는 건축의 본질에 대한 궁금증을 유쾌한 방식으로 탐구해 나가면서도 결과물은 예사롭지 않은 완성도 높은 수작으로 그 역량을 보여주었다. 이승환, 전보림(아이디알건축사사무소 대표)은 탄탄한 실무를 바탕으로 공공건축물의 시공과 절차에 대해, 전자는 BIM의 활용과 디테일한 도서로써 그 완성도를 이끌어내었고, 후자는 건축가의 실천적 과제로 제시함으로써 앞으로의 행보를 기대하게 하는 팀이었다.

7개 팀 모두 건축의 지향점과 그 해결 방식이 특별했고, 결과물의 완성도 역시 매우 우수해 심사로 우열을 가리기 어려웠다. 많은 토론이 오고간 후 김이홍과 남정민 그리고 문주호, 임지환, 조성현을 2018 젊은건축가상 수상자로 결정했다. 특히 문주호, 임지환, 조성현은 건축 전문 직능에서 소수가 알고 있는 정보를 공개하고 공유함으로써 건축의 저변이 확대되는 결과를 지향하는 새로운 접근 방식을 선보이고, 동시에 자동화된 프로그램으로 건축적 창의성이 도전받는 경계적 상황과 그 가능성을 실험해 올해 주목할 만한 건축가로 선정되었다.

공개심사에서는 젊은건축가상에 도전하는 의미에 대해 참가자 전원의 자유로운 토론이 있었다. 참가자들은 자신들의 작업 방향에 대한 의미 부여와 앞으로의 추진력으로서 수상이 강한 동기부여가 된다는 의견을 공유했다. 젊은건축가상은, 독립하여 일을 시작한 건축가가 각자의 위치에서 사유와 가치, 프로세스와 결과물에 대해 공공에게 발언하고, 접점을 공유하고 다름을 새롭게 인식하며, 앞으로의 가능성으로 발현되기를 기대하는 것으로서 의미가 있을 것이다. 또한 젊은건축가상은 여느 다른 건축상들과 달리 소위 작품이 아닌 건축가를 대상으로 한다는 점에서 그 특별함이 있다. 평가 대상은 수상 이전이지만 노련함과 원숙이 아닌 도전이 있는 젊은이라는 특정 시간대 사람에게 부여하는 상이기에 수상 이후의 시간적 지속가능성에 대한 무게감이 있다.

올해는 그 어느 해보다 많은 지원자가 있었고 작업의 결과물도 다양했으며 완성도도 매우 높았다. 우리의 도시, 경제와 문화의 변화와 맞물려 건축가의 작업량과 저변이 확대되고 있음의 신호로 여겨지며 이를 세 가지 주요한 흐름으로 읽어본다.

대규모 도시개발에서 지역과 도시재생으로의 정책 변화는 개발과 보전의 복합적인 도시적 상황을 만들어내었다. 그 결과 소규모 건축물 설계가 젊은 건축가의 새로운 도전적 과제로 이어짐을 목격하고 있다. 특히 최근의 지역과 밀착한 소규모 공공건축물 건설을 통한 지역재생 정책에서 많은 젊은 건축가의 활동과 에너지를 발견한다.

경제 규모의 발전에 따라 건축문화에 대한 성숙한 민간 건축주 세대의 증가로 민간 소규모 건축 시장의 저변이 확대되고 있다. 용적과 이윤추구만을 위한 건물이 아닌, 건축가가 만들어내는 창의적인 작업이 한 가족의 삶터에서부터 임대사업자의 부동산의 가치를 올리는 일까지 이어진다. 이는 소셜네트워크 등 매체의 다변화로 더욱 대중적으로 확장된다.

건축학인증제에 따른 5년제 교육을 받은 젊은 건축가들이 독립하여 자신의 작업을 시작하고 있다. 유학 후 독립한 기성 건축가들의 대학으로의 이동과 이들을 통한 설계교육은 유학하지 않고도 독립하는 새로운 젊은 세대로 이어지고 있다. 여전히 5년제를 졸업하고도 건축사 시험으로 자격을 취득해야 하는 제도의 불합리함이 남아 있음에도 독립하는 연령층이 더욱 젊어지고 있다.

앞으로의 더 많은 가능성은 곧 치열한 경쟁이라는 양면성으로 인식되지만, 그럼에도 젊은 건축가들로부터 각자의 위치에서 시대를 바라보는 도시 건축적 가치관에 대한 고민과 그 실천이 건축의 영역 그 이상의 범위에까지 날카로운 질문들로 도출되기를 기대한다. 이에 올해로 11회를 맞이하는 젊은건축가상은 건축가가 자신의 생각을 드러내는 공론의 장이자 동시에 이를 건축과 도시에 대한 담론으로 확장하고, 또한 동시대의 건축문화의 흐름을 읽을 수 있는 기록이자 가능성의 플랫폼 역할로 거듭나기를 기대한다.

EPILOGUE

2018 Korean Young Architect Award Evaluation Review

by Eunkyung Lee (Principal, EMA Architects & Associates)

In 2018 Korean Young Architect Award, a total of 31 teams had participated, among which only seven teams passed onto the open examination stage, which were then finally reduced to the three prize-winning teams.

Through his ability to traverse across different cultural contexts, conceptualize his observations of the given things, and produce a detailed and high-quality result, Leehong Kim's (professor, Hongik University) tenacity as he went back and forth in the boundary between concept and building in his work process was especially noticeable. Hyo Young Kim (principal, KHY Architects) showcased an impressive visual representation of order and geometry through his reinterpretation of the architectural historicity in the contemporary time and regional location. By building his continuous observation on the gap space and the boundary surface between public and private by experimenting with a factory-produced unit, Jungmin Nam (professor, Seoul National University of Science and Technology) displayed a sense of unity in his multiple project variations that begin from a small scale to the architectural space while confirming his future potential for growth. Jooho Moon, Jihwan Lim, Sunghyeon Cho (Boundless Architects) displayed a persuasive solution based on a rational analysis of the situation and the conditions in relation to the generic-style buildings built under real estate development logics and its resulting product with a high level of completeness. The part where they proposed the basic information on building construction as a widely shared methodology via a data-based automation program was especially memorable. The housing-focused portfolio of Hae-Sun Park and Seung Hyun Oh (principal, SEOGA Architects) showed how a work of young and capable architects in the general housing market can take up an influential and leading role. Hanjin Yoon, Seung Jae Han, Yang Kyu Han (principal, FHHH FRIENDS) showcased their talents through a high-quality product that resulted from their humorous method of exploring the essence of architecture. With their robust amount of work experience, this promising team composed of Seung Hwan Lee and Borim Jun (principal, IDR Architects) brought forth a high level of finish in terms of public architecture construction by utilizing BIM and a detailed picture while proposing the process of public architecture as a practical responsibility of the architect.

All seven teams had their unique viewpoints and solutions, and it was difficult to rank them as their submissions were all very exceptional. After much deliberation, we chose Leehong Kim, Jungmin Nam, and Jooho Moon, Jihwan Lim, Sunghyeon Cho as the winners of the 2018 Korean Young Architect Award. Because of their display of a new approach that aims at expanding the base of architecture by publicizing and sharing information that were only available to the minority within the professional architectural field, and their experiments on the potential and the boundary situations that challenge architectural creativity with their automated program, Jooho Moon, Jihwan Lim, Sunghyeon Cho were highlighted as the architects of the year.

At the open examination stage, a free discussion among all participants on what it means to participate in the Korean Young Architect Award took place. The participants located their reasons from their work directions and shared opinions on how being awarded would act as a strong motivation for them to continue their work.

The Korean Young Architect Award is meaningful in that it leads new architects to introduce their thoughts, values, processes, and results from their own respective worlds to the public audience, share and recognize the things that are common and different, and draw expectations towards their future potentials. Also, different from other architectural awards, the Korean Young Architect Award is special in that it does not award the work itself but the architect. Because the targets of this award are not experienced individuals but passionate young architects of that specific time, the award also comes as a weight of responsibility to the winners to continue developing themselves.

The number of participants this year was more than the previous years, and the submissions were diverse and of high quality. Following changes in the urban economy and culture, we see this as a sign that the workload and the base of architects are being expanded, and we read these three important things from this trend.

The policy change from large-scale urban development to regional and urban revitalization created a complex urban situation between development and preservation, and as a result, we notice that the young architects are engaging with small-scale architecture planning as their new tasks. We find especially much activity and energy of these young architects being recently involved in regional revitalization policy through small-scale public architecture in such regions.

With growth in the economic scale, and with the rise of building owners who are familiar with the architectural culture, the market base for private small-scale architecture is expanding. Buildings that are not merely about size or profitability but of architectural creativity are raising the values of family residences and the real estate prices of rental business operators, and this is being spread further to the public through changing forms of media such as social networks.

Young architects who have finished five years of education according to the architecture certification are beginning their independent works. The previous generations of architects who became independent after their studies abroad came back to teach, and in turn their teachings have contributed to the growth of young generations who find themselves able to become independent without needing to go abroad. Although the unreasonable system where one must study for five years and then also pass the architect exam to get licensed still exists, the age group of independent architects is getting younger.

While increasing possibilities can also doubly mean increasing competition, it is nonetheless hoped that these young architects' contemplations on the urban architectural perspective in relation to the times and their practices in their respective places would develop into acute questions to architecture and beyond. As it celebrates its 11th anniversary this year, it is also hoped that the Korean Young Architect Award will be revived as a place where architects can display their thoughts publicly, expand their thoughts into discussions on architecture and the city, and be established as a platform that embodies the flow of contemporary architectural culture through its functions as a record of the past and a springboard for the future.

도판 출처
IMAGE CREDITS

p. 16 ©designboom (www.designboom.com)

p. 18 *Home within Home within Home within Home within Home*, 2013
Site-specific commissioned artwork for Hanjin Shipping
Box Project, MMCA (National Museum of Modern and
Contemporary Art), Seoul, 13 November 2013 – 11 May 2014
©Do Ho Suh, Courtesy of the artist, Lehmann Maupin New
York, Hong Kong and Seoul and MMCA (National Museum of
Modern and Contemporary Art)

p. 20 ©Will Pryce

p. 22 ©Matthew Monteith

p. 24 ©Leehong Kim

p. 26 Escher, Relativity, ©The M.C. Escher Company B.V.

p. 28 ©Kiduck Kim

p. 30 ©Porsche (www.porsche.com)

p. 32 ©Steven Holl Architects

p. 118 ©Jayeon Kim

p. 120 ©Jan Vormann

p. 122 Google Earth

p. 216 ©Jungmin Nam

p. 218 ©Jungmin Nam

p. 220 ©Federica Violin

p. 224 Teruobu Fujimori, Dandelion House ©Iino Mitsuru

p. 226 ©Rick Olmstead

p. 228 Farshid Moussavi and Michael Kubo, *The Function of Ornament*,
Actar (January 15, 2006)

p. 230 Robert Venturi, Steven Izenour, and Denise Scott Brown,
Learning from Las Vegas, The MIT Press; revised edition
(June 15, 1977)

p. 232 ©Guitar photographer